Traditional British Poultry Breeds

CANCELLED

Traditional British Poultry Breeds

BENJAMIN CROSBY

THE CROWOOD PRESS

First published in 2012 by
The Crowood Press Ltd
Ramsbury, Marlborough
Wiltshire SN8 2HR

www.crowood.com

© Benjamin Crosby 2012

All rights reserved. No part of this publication may be reproduced or transmitted in any form or by any means, electronic or mechanical, including photocopy, recording, or any information storage and retrieval system, without permission in writing from the publishers.

British Library Cataloguing-in-Publication Data
A catalogue record for this book is available from the British Library.

ISBN 978 1 84797 337 5

Dedication
This guide is dedicated to those who do more for others than they do for themselves. In my case many people have helped me along the way, and these are duly recognized in the acknowledgements of this guide. Many people are dedicated to the continued existence and promotion of traditional British breeds. This guide is also dedicated to the efforts of those people.

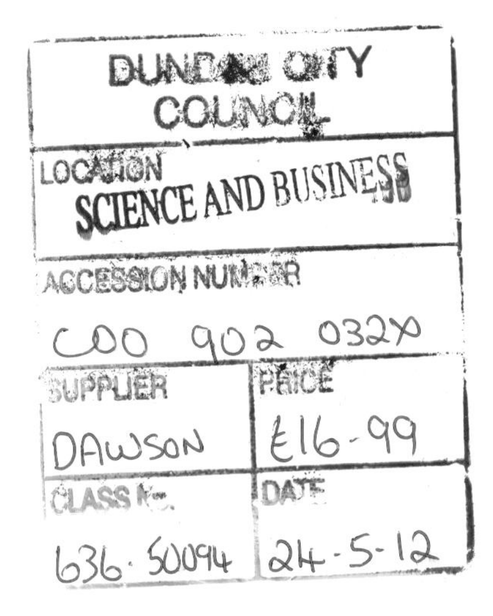

Typeset by Jean Cussons Typesetting, Diss, Norfolk
Printed and bound in Malaysia by Times Offset (M) Sdn Bhd

Contents

Acknowledgements

First and foremost I thank my parents, Joyce and Trevor Crosby, for their understanding when I converted their back garden to accommodate my poultry in the early years, and for their support in helping me to look after my flock at difficult times (of which there have been many). I also thank my grandparents, Daphne and Stanley Wilson, for supporting my interest in native breeds, and for their opinions.

I also thank Gavin and Irene Keys, long-term breeders of Buff Orpingtons, for their good ideas, support and encouragement to make this book and many other things possible; Andrew Gemmill and family of New Hockley Hall Farm, Essex, for allowing me to use land at their farm to expand and improve my collection of native breeds, and for their useful advice; Clive Greatorex of Ramsey Forty Foot, Cambridgeshire, a good friend and past breeder of Suffolk Punch horses and Marsh Daisy and Scots Dumpy poultry (Clive has been particularly helpful, especially in the search for good quality Marsh Daisy stock); Charlie Peck of South Godstone, Surrey, for his help with the qualities of the Marsh Daisy breed – Charlie is a great fan of utility poultry breeds, and has been an inspiration; and finally Alex Bull and family, a good friend who has supported my interest in British breeds for many years. Being a web developer, Alex has been hugely influential in the development of several useful websites, including www.poultryshop.co.uk, www.britannicrarebreeds.co.uk and www.homesufficiency.co.uk

My thanks also go to the following people for their help: Andrew John Macredie of Belph Grange Farm, Worksop, Nottinghamshire, for supplying photographs of Orpingtons; Andrew Sheppy of Congresbury, Bristol, for information about his Gold Brussbars; Chris and Mike Ashton and their in-depth and accurate guides to ducks and geese (also published by The Crowood Press); Christine Taylor of Market Drayton, Shropshire, for supplying me with my first Bantam Derbyshire Redcaps, for information on the Bantam Redcap, and for general advice; David Scrivener for his help with breed histories and other useful advice (David's guides *Rare Poultry Breeds* and *Popular Poultry Breeds*, also published by The Crowood Press, have been very helpful); Frank Bridgland for his help with the history of the Norfolk Grey (I also admire his keen determination to get the existing Norfolk Grey recognized by the RBST as a rare breed); Jane Eardley of Buxton, Derbyshire, for supplying detailed photographs of Rosecomb bantams; Kathleen Arnold of Philcott Poultry, Bury St Edmunds, for supplying photographs of Gold and Silver Wybars; Martin Baldwin of Tryal Farm for supplying photographs of Welsh Harlequin ducks; Mary Isbister of Trondra, Shetland, for helping with the breed history of the Shetland duck and Shetland geese and for photos of Shetland geese; Peter Trumpeter and associates of Burtons Green, Essex, breeders of Ixworths and many native pigs (Peter has been inspirational and offered practical ideas along the way); Robin Creighton of Great Leighs, Essex for helping me appreciate the qualities of the Sebright breed, for letting me take photos of his Sebrights, and for supplying a photo of a Silver Spangled Hamburgh; Simon Mckean of Deighton, Huddersfield for helping me understand the qualities of the Modern Game and for supplying photographs of various colours of the breed; Steve and Liz Bateman of Chesham, Buckinghamshire, breeders of Ixworths and Marsh Daisies, among other British breeds (Steve and Liz are living the dream at Hazeldene Farm and are very encouraging every time I visit); Sue Feild of Victoria's Poultry, Chester, for photos of Coronation Sussex; Trevor Martin for his help with the history and qualities of the Suffolk Chequer, for supplying me with my first Suffolk Chequers, and for lots of other helpful advice; Tony Bennett of Ramsbottom, Lancashire, for his advice on auto-sexing breeds; and William Osborne of Lydney, Gloucestershire, for his past help with the history of the Stanbridge White and Norfolk Grey.

All poultry and photographs are the author's unless otherwise stated.

Introduction

This practical guide will make interesting reading for hobbyists and home owners wishing to live a little more self-sufficiently, and for conservationists and agricultural students who wish to clearly understand and appreciate what breeds exist, how they can be utilized and managed, and how to identify good quality stock.

Most British breeds have been forgotten by hobbyists due to the influx of fashionable foreign breeds into the fancy and the continued development of a select few breeds for showing. This guide will help make British breeds once again known to the hobbyist, and aims to describe, as fairly as possible, the attributes of each breed, helping the hobbyist determine which breed or breeds might be suitable for their own circumstances.

This guide is perhaps like no other in that it gathers together practical information on all core British breeds in one resource, and contains a vast photographic collection of native breeds. Unlike other guides that illustrate the near feather-perfect show specimen, this practical guide shows common breeding stock and the variations that may sometimes occur. Photographs of common flaws and useful or interesting features are also included to make it clear what should be avoided or selected for.

BREEDS INCLUDED IN THIS GUIDE

Domestic Fowl A–Z

Dorking
Carlisle Old English Game
Hamburgh
Indian Game
Ixworth
Lincolnshire Buff
Marsh Daisy
Modern Game
Norfolk Grey
Old English Pheasant Fowl
Oxford Old English Game
Orpington and Australorp
Rosecomb
Scots Dumpy
Scots Grey
Sebright
Suffolk Chequer
Sussex

Auto-sexing Breeds

Brussbar
Legbar
Rhodebar
Welbar
Wybar

Ducks

Abacot Ranger
Aylesbury
Campbell
Magpie
Orpington
Shetland
Silver Appleyard
Stanbridge White
Welsh Harlequin

Geese

Brecon Buff
Shetland
West of England

1 British Poultry Breeds: History and Definition

HISTORY OF DOMESTIC FOWL

Domesticated poultry derived from jungle fowl, and the correct term to describe them is 'domestic fowl': this is equivalent to the commonly used term 'chicken'. The oldest British poultry breeds are the Dorking and Old English Game; these breeds are truly ancient, and are known to have existed in Great Britain for more than 2,000 years. However, their history is largely unknown. The Dorking was traditionally used as a table bird, and the Old English Game, previously known as Game Fowl, was used for the social diversion of cock fighting. These are the two main reasons why poultry were originally kept in Great Britain.

Even though domesticated fowl have existed in Great Britain for over 2,000 years their ancestors would have been imported into the country. However, because there is little mention of these breeds in early records, it is not known how they evolved over the first 1,800 years or more since arriving in Great Britain. The Dorking and Old English Game are included in this guide because they have existed in this country for several centuries, during which time they are likely to have been frequently crossed and selected to suit the needs of British keepers.

There are several other poultry breeds besides the Dorking and Old English Game that have long existed in Britain, whose development may at least in part have involved the use of the Dorking or Old English Game. These breeds include the Derbyshire Redcap, the Old English Pheasant Fowl, the Hamburgh, Scots Grey and Scots Dumpy. The early development of these breeds is completely unknown, but they were originally localized farmyard fowl that were later named and standardized to comply with the exhibition scene following the first national poultry show held at London Zoo in 1845.

With the increased demand for exotic breeds following this first national show, many Asian and Mediterranean breeds were imported into Great Britain and these breeds have been hugely influential in the development of productive British poultry breeds. The Asian influence included the Brahma, a breed developed from Asian stock in America, first imported in 1853 and used in the breeding of the Sussex. The Cochin came from the Far East in 1847 and was used in the development of some Orpington colours and the Lincolnshire Buff. The Langshan was imported from China in the early nineteenth century; it was first shown in 1872 by Major F. T. Croad. These breeds were large-framed birds that laid brown eggs, and set the trend for pale brown eggs in Great Britain. Until their introduction, British poultry breeds produced white or tinted eggs.

The Mediterranean introductions included the Ancona, Leghorn (introduced in 1869) and Minorca. These Mediterranean birds were prolific layers and were used by British breeders in an attempt to establish productive commercial laying breeds. Many of those influenced by the Asian and Mediterranean introductions were developed between 1900 and 1940, and include the Ixworth, Marsh

Daisy, Norfolk Grey and Orpington. The Orpington was the first to be specifically and intentionally developed as a dual-purpose breed, by William Cook, and although it was quickly developed as an exhibition bird, its utility merit as a good layer and table bird lives on in the Australorp.

The Sussex breed took longer to develop, and was standardized after the Orpington. The Norfolk Grey and Ixworth were established later, and were also developed to be dual-purpose commercial birds. Between 1903, when the Sussex was first standardized, and the 1960s when hybrids took control of the commercial poultry industry, the Light Sussex was one of the most widely used commercial poultry breeds in Great Britain, and could probably be found in greater numbers than all the other British poultry breeds put together. Introduced breeds also used in abundance were the Rhode Island Red, Leghorn and North Holland Blue.

The auto-sexing breeds were the next step in the evolution of British breeds. They were developed along scientific lines by Professor Punnett and Mr Peace at the Cambridge University Agricultural Research Department. Their work took place from 1922 onwards at a time when genetics, and in particular poultry genetics, were largely unknown. Many of the auto-sexing breeds were originally developed to be commercially viable alternatives to existing pure breeds. Unfortunately for pure breeds, the hybrid strains developed after World War II had hybrid vigour, and were carefully selected with a view to maximum production for minimal feed consumption. As a result hybrid birds became specialized for maximum output of either eggs or meat as broilers. Much of this hybrid development occurred in the USA.

By the 1960s standardized pure breeds were no longer commercially viable for the mainstream poultry industry, and since then have been maintained almost exclusively by hobbyists; indeed many of these breeds have become quite rare. Fortunately a resurgence in interest in poultry keeping has meant that many of the traditional breeds described in this guide have become a practical and decorative option for today's relaxed and enthusiastic hobbyist. This guide will describe the great variety of existing native breeds, and will show that there are useful, attractive and challenging alternatives to common favourites.

Light Sussex hen.

History of Domestic Waterfowl

All native ducks are derived from the Mallard (*Anas platyrhynchos*), and native domestic geese originate from wild Greylag geese (*Anser anser*). Aylesbury ducks and Shetland and West of England geese are the earliest existing British waterfowl breeds, and were gradually developed over many centuries to suit the needs of farmers and crofters. However, most existing British duck breeds were specifically developed in the first half of the twentieth century to meet the demand for birds that were more productive and commercially viable, and

CHANGING TRENDS

Trends in the fancy are often colour orientated, but there is also an apparent preference for profusely feathered and ornamental breeds such as the Cochin, Faverolle, Orpington, Pekin and Poland. Game Fowl still have a keen but somewhat secretive following.

Colour is not the only characteristic affected by trends: the size of the bird has also become an important factor. In the case of many breeds, bantams are now equally or indeed more popular than their large fowl counterparts. Many hobbyists now choose to keep bantams as an attractive and easy-to-manage pet, since they require less feed and less space than large fowl breeds. From the popularity of certain breeds at shows and auctions it would appear that many hobbyists prefer to choose a breed for its colour and size rather than for its potential usefulness.

The Hamburgh, Orpington and Sussex are among the most popular of all British poultry breeds, and are commonly seen in bantam form. Some breeds do not have a bantam counterpart at all; others are more popular as large fowl than bantam, and these include the Derbyshire Redcap, Dorking, Indian Game, Rhodebar, Scots Dumpy, Scots Grey and Welbar. Unlike the Sussex and Orpington, many of these breeds are quite rare, and the bantam versions even more so. Some of the bantam forms have a wild or flighty habit, which perhaps deters hobbyists from keeping them, and may be the reason that some of these breeds have not followed the trend for bantams.

often of an attractive colour. Indeed, one of the reasons for this explosion of development was the introduction of the Indian Runner in the late 1870s. Not only was the Runner recognized as a highly productive layer, but it also came in many colours, so breeders could develop duck breeds in colours other than white, black or Mallard colour.

Definition of a Rare Breed

Officially a rare breed of poultry is one that does not have its own breed club, but realistically even the many breeds that do have a breed club may still be classed as rare, and mainly restricted to localized areas. Often a breed itself is not rare, but certain colour varieties are extremely rare. This is the case with the Sussex, where the light and buff varieties are very common, yet the brown, coronation, red and white varieties together could not match the number of the Light Sussex variety alone.

Furthermore the problems associated with some of the breeds or varieties described in this guide may give the impression that certain British breeds are difficult or unreliable to breed and maintain. This is often why they are rare, but this does not necessarily rule them out as unsuitable for devoted hobbyists – and it is important to remember that even introduced breeds have their problems and variations.

In general there is far less interest in waterfowl than there is in domestic fowl, and with the exception of perhaps the Campbell duck, all British duck and goose breeds are in short supply; furthermore some vary greatly in quality and purity. Shetland, Silver Bantam and Stanbridge White are probably the rarest of the duck breeds, while the Aylesbury and Silver Appleyard vary the most in quality and purity.

WHAT IS A BRITISH POULTRY BREED?

This is a topic which is likely to cause some controversy in the fancy because there are many grey areas that need to be addressed;

Popular and common breeds	Rare breeds (with or without a breed club)
Dorking	Brussbar
Hamburgh	Derbyshire Redcap
Indian Game	Ixworth
Modern Game	Legbar
Old English Game Bantam	Lincolnshire Buff
Carlisle Old English Game	Marsh Daisy
Oxford Old English Game	Norfolk Grey
Orpington (The Australorp is a very rare standardized breed in Great Britain, but is effectively a version of the Black Orpington)	Old English Pheasant Fowl
Rosecomb	Rhodebar
Sebright	Scots Dumpy
Sussex	Scots Grey
	Suffolk Chequer
	Welbar
	Wybar

furthermore a list of core breeds is needed to appreciate which breeds can be truly regarded as native.

The term 'British' is mainly used to refer to Great Britain; however, it can also refer to the United Kingdom. For the purpose of this guide the term 'British' will refer to poultry breeds that originate from the island of Great Britain, defined as including England, Scotland and Wales.

So how do we define 'breed'? A breed is a group of birds within a species that have been deliberately created to have a distinctive appearance or attributes. The origin of a breed is determined by the place in which the original breed (standardized or not) was created. This guide deals only with the core British poultry breeds, those that are undoubtedly of British origin, as opposed to those that are perhaps merely British versions of established foreign breeds.

West of England geese.

Selection alone of an established foreign breed, even if it changes the type, size or characteristics of a breed, is not enough to change its nationality: thus a new selection in Great Britain would merely be a British version of the original breed. In Great Britain some crossing with an unrelated breed is necessary, but to be classed as British the resulting stock must be developed into a breed that is unique from the original stock.

Breeds that have existed in this country before records on the breed began, and which do not have any identical foreign counterpart, can be classed as native because their developmental history is largely unknown

– in essence the breed is given the benefit of the doubt, and an assumption is made that the breed has been in Britain long enough to have been crossed, or for it to have evolved substantially into a unique breed.

If a breed is crossed to enhance a feature such as the size of the ear lobes or to create a new colour, as long as it retains its established type and key characteristics then it remains merely a new variety of an established breed. For example this is the case with new colour varieties of Orpington: thus if a breeder introduces Wyandotte to the Orpington to create a laced colour variety, yet maintains the key type and characteristics of the Orpington, including the name, then it is purely a new variety of the Orpington breed irrespective of where in the world the new variety is made.

Breeds that some hobbyists may consider as being British but are not included in this guide of core British breeds, include the following: Ancona, Andalusian, Burmese Bantam, Campine, Croad Langshan, Minorca, Modern Langshan, Shetland (non-standardized domestic fowl similar to *Araucana*), Spanish.

Why is Origin Important?

In the first instance it is only fair that the breeder, and the location where a breed is created, are given credit for their achievements. Origin was particularly important in the late nineteenth and early twentieth centuries, at a time when poultry breeding and the demand for ever-more productive breeds was commercially important. Even a farmer with limited resources could influence the poultry world.

Historically it is useful to know the origin of breeds, as this can show how poultry breeders, fashions, ideas and locations have influenced the evolution of poultry. Today, origin is important to breeders wishing to maintain a breed that is local to them, or which has some connection with their family members or family history.

By having a list of core British breeds it will be possible for hobbyists to appreciate which breeds were developed in Great Britain, and will also clearly show them which native breeds could be used or preserved.

Variation in Poultry Breeds

Different strains develop within a breed, and these may vary greatly depending on breeding, management and selection. Some strains are the result of specialist selection for their merit as layers or table birds, others are bred for exhibition purposes, while many are simply 'run-of-the-mill' stock. Most breeds are not nearly as productive as their ancestors, and hobbyists generally have neither the time nor the facilities to accurately select and improve birds for their performance abilities. Moreover breeders often choose to take the easy route by reintroducing one of the parent breeds, or

Buff Orpington drake.

by introducing a new breed or hybrid to get quicker results – though this is ultimately to the detriment of the original breed.

This sort of 'improvement' could be considered an evolution of the existing breed, encouraging hybrid vigour and improved productivity – or it may cause the breed to degenerate in quality and lead to a lack of consistency in the breed as a whole. In particular this may be the case where a breed is manipulated to suit a short-term fashion, a trend which is typical of twenty-first-century society as a whole. Degeneration in quality and lack of consistency can already be seen in the Marsh Daisy breed, due to their rarity, and is beginning to manifest itself in Silver Appleyards, Indian Game, Norfolk Greys, some of the rare colours of the Sussex, the size of Hamburghs and the hardiness and productivity of Dorkings.

It is sometimes difficult to fully appreciate the true qualities of a breed because variation will sometimes exist legitimately within it. Variations in habit can occur, including broodiness, temperament, flightiness and hardiness, which means that different breeders may have differing opinions as to the qualities of a breed. However, there is a generally accepted set of attributes for each breed, often only appreciated by a breeder once he has sourced birds from many outlets and maintained them for several years. Thus a variation in habit could be due to selection, or outcrossing with another breed.

A good example of this is the Derbyshire Redcap: the large fowl Redcap is universally recognized as being a poor sitter, yet the bantam version that looks identical except in size is known to be quite reliable. The reason for this is that the Wyandotte, a known sitter, was used to help establish the bantam version of the Redcap. During this process such attributes as broodiness were not selected against, because the focus was on making the bantam version appear the same as the large fowl, but smaller. This example illustrates that such an outcross, albeit well in the past, can still have an effect on the current habit or productivity of a breed, even if it does not appear to have affected its appearance.

The hobbyist must also appreciate that no British poultry breed can truly fly, and all native breeds are susceptible to predators such as the fox. However, some lightweight breeds do 'push the boundaries' of this assumption, and are flighty and more evasive of predators than heavyweight breeds.

The existence of incorrect and inferior stock in the marketplace, and the rarity of some breeds, sometimes means that the hobbyist has the difficult, frustrating and expensive task of buying birds from many sources, breeding them, and then culling those that don't meet the grade, before they arrive at the correct, required standard. Furthermore, sourcing birds from exhibitors is by no means a guarantee of good results because often exhibition strains will have poor productivity and incorrect egg colour, as they are selected primarily for their external appearance and are sometimes highly inbred.

Outcrossing has been going on for a long time, either accidentally and unknown to the breeder, or deliberately, so apart from a few long-established closed flocks there is really no such a thing as a pure breed. However, as long as the breed standard is adhered to, the bird will superficially look as it should. Many rare breeds and colour varieties have come close to extinction several times and are often unknown to experienced poultry keepers. Unfortunately this means that hobbyists do not have the information, photos or easily available advice to ensure that birds are correctly selected and bred to maintain their quality.

CHOOSING A SPECIES AND A BREED

Domestic Fowl or Waterfowl?

As a general rule domestic waterfowl are hardier, longer lived and more productive than domestic fowl. Duck breeds such as the Campbell produce more eggs than any chicken breed, and all ducks and geese grow faster, making a viable table bird at a younger age than domestic fowl. So why are chickens more popular? There are several reasons: first, there is a great deal more variety among chicken breeds, including differences in utility merit, plumage colour, egg colour and physical features. Chicken eggs are easier to crack, and chickens pluck more easily and quickly.

There is a common belief that it is essential to have a pond if you intend keeping ducks and geese, and this often deters people from keeping them. However, although they would maintain themselves in better condition with a pond, it is not essential. Instead, a small paddling pool filled with fresh water every now and again would allow them to clean and oil their feathers thoroughly. Day to day, a bucket of water or an automatic drinker would suffice.

Even so, ducks and geese still require more intensive management because of their messy habit. They are aggressive grazers and like to dib their beak in soft ground in search of food, and the soil around their beak then usually ends up in the drinking water. Similarly, if they are fed meal or mash a small amount usually ends up stuck to their beak, which they will rinse off in the drinking water. Both soil and feed quickly foul fresh water. Furthermore their wet droppings soon saturate bedding material, necessitating more frequent cleaning out.

Nevertheless chickens, ducks and geese potentially have a part to play for every hobbyist and smallholder. It is up to you to decide which breed suits your requirements, and the environment you have available.

Traditional Breeds

As will be described later, some native breeds are hard to come by, and due to their rarity are sometimes of poor or variable quality – they may lack vigour or hardiness and may not have the productive capabilities they once had. Fortunately most native breeds are still productive enough to be worthwhile maintaining, and with patience they can be improved further.

Traditional breeds also have the advantage of coming in a wide range of attractive varieties. In some cases they have been developed for hardy free-range rearing, and have been challenged to survive in a harsh outdoor environment, often without adequate protection.

Traditional breeds can also be used as foundation stock in the creation of new hybrid varieties suited to the changing needs of hobbyists. Light Sussex females can be crossed with Gold Silkie males to create sex-linked broodies with reliable mothering instincts; these hybrids are known affectionately as 'Gold Tops'.

2 The Uses of British Poultry Breeds

THE USES OF POULTRY

People keep poultry for a whole host of different reasons, probably in the main for egg production and as table birds, but their feathers and down and even their manure all have a use, geese and ducks are exploited for their grazing ability, and geese have even been used as 'guard dogs'!

Geese as Guardians

Geese are widely recognized as extremely vocal birds, and will raise the alarm by making a considerable noise when disturbed, particularly by strangers, even late at night. Geese were sometimes used instead of dogs at police stations during the troubles in Northern Ireland, as dogs were more easily distracted by food.

Poultry for Egg Production

Probably the most important use of poultry, and the main reason that most hobbyists keep a few chickens, is to produce eggs. Egg colour varies greatly among British breeds, from white, through various tinted shades to pale brown, medium brown, pale blue and pale green. Egg-laying ability varies hugely between breeds and will also depend on the strain; thus exhibition strains are usually regarded as being less productive layers than utility strains – although reliable utility strains are few and far between. Most hobbyists have fairly ordinary stock, but with the potential for improvement in either direction.

Eggs.

Poultry for Exhibition Purposes

Poultry exhibiting is a large and long-established part of the domestic poultry

Show cage.

industry, and originally of the commercial poultry industry too. There are many British breeds that have been specifically developed or refined for exhibiting; a few of these are the Modern Game, Orpington, Rosecomb, Sebright, Suffolk Chequer, Aylesbury and Magpie.

Duck Down and Goose Quills

Duck feathers and down are used in duvets and pillows. White breeds such as the Aylesbury, White Campbell or Stanbridge White are most suitable for providing clean-looking, consistent pillow contents.

The primary wing feathers from geese are traditionally used as quills for writing.

Duck-down pillow contents.

Waterfowl for Grazing and Agro-forestry

Ducks and geese are very aggressive grazers, and given enough room, or worked in rotation, they can be effectively employed to keep grass and weeds down and control insects in orchards, vineyards or plantations, particularly in the early years of a plantation's development. At the same time they will provide eggs, and with additional feed will grow on and finish as table birds.

Poultry are more suitable in these

Free-range Derbyshire Redcaps.

circumstances than sheep, pigs or cattle, which would strip and damage bark or eat the young shoots on plants, therefore requiring each plant to have protection. Poultry will only eat leaves that are in reach. Recently planted whips or young pot-grown plants may need individual protection, as ducks and geese can damage the bark on young plants. Otherwise vines and trees can easily be grown or trained to be out of reach of poultry.

Ducks and geese are less likely to escape than chickens, but would require a large water source, perhaps an automatic drinker connected to a barrel or the mains water supply. A pond is not essential, although it would be beneficial.

Poultry in such an agro-forestry system require a secure pen enclosing the area to protect them against predators. The easiest and least permanent method would be electric mesh fencing. Chickens are less suited to this management method as they are more likely to jump the mesh, and may even roost in the trees of an orchard or perch on vines and posts and wire in a vineyard, making it more difficult to control them and to shut them in at night if necessary.

Poultry as Table Birds

The breeds that formed the mainstay of the table-bird market in London during the late nineteenth and early twentieth centuries included Sussex and Lincolnshire Buff chickens and Aylesbury ducks; today the Sussex can still be used by the hobbyist as a hardy and attractive table bird, although they would always be far from achieving the prolific weight gain of modern-day broilers.

Other potentially useful table breeds include the Australorp, the Indian Game and the Ixworth. There are many other British breeds which, although far less productive, were considered dual-purpose breeds and may have been originally used as table fowl by their owners – for example the Derbyshire Redcap and the Norfolk Grey.

Ixworth and White Sussex free ranging on a sheltered, grassy pen at Hazeldene Farm, Chesham, Bucks.

The Benefits of Poultry Manure

Well rotted manure is an excellent way of improving soil structure and will help to retain moisture; poultry manure is also high in nitrogen. It is best added to the compost heap with straw and/or other plant material to prevent the compost slumping, and to ensure a fibrous texture. If fresh poultry manure is fed directly to existing plants it is likely to harm the roots and may attract vermin, as well as looking and smelling bad.

Poultry manure can be spread and ploughed directly into farmland soil in the autumn before planting takes place in spring; in a garden environment it is best mixed with other materials and left to compost first.

SPECIALIST BREEDS

Productive Laying Breeds

A specialist egg-laying breed should be an efficient and prolific producer of eggs for a minimal input of feed. It is not necessary to have a large bird in order to have eggs of a large size or in large quantity; even lightweight and small breeds have the potential to produce large eggs in relation to their size.

Egg colour is commercially important: traditionally in Great Britain the public prefer brown eggs, and nowadays both dark brown and blue eggs are fashionable, while in the USA they prefer white. Thus in order to meet market demand in Great Britain, it is commercially important to choose a breed that lays brown eggs.

Brooding instincts are not desirable in layers because productivity will drop while the bird is brooding, and increased management is required to prevent broodiness and to remove brooding hens.

Profuse feathering may be a hindrance as the feathers can become congested with dirt, particularly if the bird is messy or clumsy with eggs. A lightweight, energetic bird is preferred with reasonably tight feathering, that matures early and feathers quickly when young.

The most suitable breeds include the following:

Domestic fowl: Australorp, Brussbar, Derbyshire Redcap, Hamburgh, Ixworth, Legbar, Marsh Daisy, Norfolk Grey, Old English Pheasant Fowl, Rhodebar, Scots Dumpy, Scots Grey, Sussex, Welbar, Wybar.

Waterfowl: Abacot Ranger, Campbell, Magpie, Stanbridge White, Welsh Harlequin.

Productive Table Breeds

Table birds of traditional breeds should develop a full breast of meat by the time they are mature at about twenty-five weeks of age – the earlier they develop, the more cost effective they become. Be aware, however, that a full breast of meat on a traditional bird is far less than that of a commercial broiler raised for the supermarket. Some traditional breeds can take more than a year to put on enough meat for it to be worthwhile using them as table birds, and some will never put on enough: therefore choosing the right breed is important.

As a result of their continuing rarity, inbreeding, and lack of focused selection by hobbyists, very few traditional breeds are worthwhile as table birds, even if the history books say they should be. Lightweight breeds such as the Old English Pheasant Fowl can also be used as table fowl, although one bird is only likely to feed two people. The meat of traditional breeds does, however, mature more slowly, thereby on the one hand improving the flavour, but also making the meat tougher, particularly the legs.

Table breeds require strong legs to support their weight, and long legs and a nearer-to-vertical back allow for easier mating – poor fertility is often a problem in some heavy table birds. A long, straight keel with a well rounded breast is ideal. The keel should be covered with feather to prevent reddening or discoloration to the skin.

A bird with white plumage, particularly on the main body and breast, was always preferred by the commercial market because in a coloured bird the dark stubs left behind after plucking were considered to be undesirable. However, this preference now appears to be reversed, as the general public are beginning to believe that dark stubs suggest the bird is more naturally reared or of a traditional type.

The most suitable breeds include the following:

Domestic fowl: Australorp, Indian Game, Ixworth, Lincolnshire Buff, Sussex.

Waterfowl: Aylesbury, Silver Appleyard, Brecon Buff.

DUAL-PURPOSE BREED

A dual-purpose breed is a highly productive egg- and meat-producer, and is ideally suited to the hobbyist and smallholder who require an all-in-one bird. Dual-purpose breeds are not normally as prolific as a specialist egg or meat breed. They are categorized as lightweight or heavyweight.

The most suitable breeds include the following:

Domestic fowl: Australorp, Brussbar, Derbyshire Redcap, Ixworth, Lincolnshire Buff, Norfolk Grey, Old English Game, Old English Pheasant Fowl, Rhodebar, Scots Dumpy, Scots Grey, Sussex, Wybar.

Waterfowl: Magpie, Orpington, Silver Appleyard, Stanbridge White, Brecon Buff.

Standard Free-Range Breeds

For general purpose, free-range birds managed on grass or scrubland it is useful to have a foraging breed that actively seeks out food to supplement its diet, although the main diet will need to be supplied. Breeds that are hardy and vigorous and not specifically developed for exhibiting are ideal. Secure fencing is required as many of these breeds cannot escape from or defend themselves against predators. Most native breeds are suitable for free-range rearing, managed in large pens, on open paddocks or in woodland.

Low Maintenance, Free-Range Breeds

Some native free-range breeds require less maintenance than others, and due to their hardiness can withstand a harsh environment and evade predators more readily than other breeds. A key attribute of these low maintenance breeds is their small appetite, and if necessary they can forage much of their own food when left on a grass meadow or wooded range at a low stocking density.

Good foragers usually have a lively carriage and are a 'gamey' type, with a lightweight body and good eyesight. They are the most self-reliant of the native breeds, and are most likely to evade predators. Good feet are important; a fifth toe is unnecessary, although not a hindrance. A small comb held close to the head, such as a rose or pea comb, is less likely to be damaged when roaming along hedgerows and through wooded areas. Short legs may hinder speed and activity, but they do help in keeping birds hidden in long grass. A dark plumage colour such as black with a glossy green sheen, or black-red, partridge or wheaten colours will successfully camouflage birds in overgrown or shaded areas.

All these attributes are important if birds are to roam over a farm or large area without confinement. Many native lightweight breeds have indeed been bred and naturally selected over many centuries to survive a low maintenance environment. Some lightweight breeds are particularly useful if a hobbyist intends to leave them to their own devices around a home or a farmyard during the day, while locking them away at night.

Despite their hardiness and active lifestyle, even these breeds are susceptible to predators such as foxes, so the hobbyist's individual circumstances and environment must be carefully considered.

The most suitable breeds for this lifestyle include the following:

Domestic fowl: Derbyshire Redcap, Hamburgh, Norfolk Grey, Old English Pheasant Fowl, Carlisle Old English Game, Oxford Old English Game, Scots Dumpy, Scots Grey.

Waterfowl: Shetland duck, Silver Bantam, Miniature Silver Appleyard, Brecon Buff, West of England, Shetland geese.

Reliable Sitters

A good sitter should be reliably broody and persistent in her attempt to hatch eggs. A large breed with profuse plumage would be suitable. Heavy breeds cover a greater number of eggs, but some, such as the Indian Game, are sometimes clumsy due to their weight and ungainly carriage, resulting in broken eggs. An aggressive nature when broody is also useful when eggs or chicks require protection from predators such as cats and rats. Old English Game and Indian Game are notably very defensive. Most duck breeds are not reliable sitters.

The most suitable breeds include the following:

Domestic fowl: Dorking, Indian Game, Ixworth, Lincolnshire Buff, Marsh Daisy, Old English Game, Orpington, Scots Dumpy, Sussex.

Broody Brown Marsh Daisy hen.

Waterfowl: Brecon Buff, Shetland, West of England.

Exhibition Breeds

Some breeds, such as the Sebright, have been developed purely for exhibition, while others have evolved from a general purpose bird to one of exhibition type; the most notable of these is the Orpington. Exhibition breeds are often exaggerated in type and plumage: for example, long legs have been bred for in breeds such as the Modern Game to create a delicate, ornamental look. Other breeds have been developed away from the gamey type, often with shorter legs, a horizontal back and bigger body, resulting in a slower carriage less suited to foraging. Plumage is often exaggerated, creating a bigger-looking bird. Feather colour and markings are perfected in exhibition stock with consistency throughout. Unfortunately exhibition strains are not known for high hatchability, hardiness or high productivity, but they often completely meet the requirements of the breed standard. Sometimes particular attributes of a breed, such as the ear lobes on the Rosecomb breed, become focal points that come under much scrutiny.

Breeds that are often used and developed specifically for exhibition include the following:

Domestic fowl: Hamburgh, Modern Game, Old English Game Bantam, Carlisle Old English Game, Orpington, Rosecomb, Sebright, Suffolk Chequer, Sussex.

Waterfowl: Aylesbury, Magpie.

Breeds for People with Limited Space

The most suitable breeds for people with limited space, or for those living in a town or city, should be small in size so they can be kept in a small aviary, fold unit (an ark with a run), ark or large cage. They can be kept for either exhibiting or for egg production. Bantams are more suited to confined management than large fowl, particularly those of non-flighty breeds.

Auto-sexing ability is not essential but would be very useful if hobbyists wish to breed from their poultry but limit the numbers kept. For egg production the cockerels are obviously not required, so the sooner they are identified by the hobbyist and culled out, the better. Most bantam versions of productive breeds are suitable, but breeds with auto-sexing ability would prove more practical. Unfortunately the auto-sexing breeds have become rare to the point that sourcing stock may be very difficult.

One other matter for consideration is the level of noise made by poultry. Hens and drakes are relatively quiet, but cockerels, cocks and ducks can be very noisy. A certain level of noise is inevitable, but it can be limited by keeping only one mature male, also by keeping birds locked away in their house until mid-morning, or by keeping them permanently in a shed.

The most suitable breeds include the following:

Domestic fowl: Rosecomb, Sebright, Suffolk Chequer, Old English Game bantam, bantam Orpington, bantam Sussex, bantam Scots Grey, bantam Rhodebar, bantam Welbar, bantam Wybar.

Waterfowl: Silver Bantam, Miniature Silver Appleyard.

3 Domestic Fowl A–Z

DERBYSHIRE REDCAP

Origin: Derbyshire and surrounding counties, England
Class: Light, soft feather
Colour varieties: Gold-spangled
Purpose: Eggs, meat, exhibition
Eggs: White, medium-sized
Weight: Large fowl cock, 2.7–2.95kg (6–6.5lb); cockerel, 2.5–2.7kg (5.5–6lb); large fowl hen, 2.25–2.5kg (5–5.5lb); pullet, 2–2.25kg (4.5–5lb)
Ancestry: Not known
Sitter: No
Auto-sexing: No

Large fowl Derbyshire Redcap cockerel.

Large fowl Derbyshire Redcap hen.

History

Year created: Not known.

Originator: Not known.

Breed development: The Redcap is a dual-purpose farmyard breed that was further selected and improved in Derbyshire specifically for its oversized rosecomb. The comb was an important feature of birds exhibited in local pub shows during the nineteenth century.

Bantam: The bantam version was created in about 1930 by Messrs Baker, John Elliott and H. O. Young by crossing undersized large

Redcaps with bantam Partridge Wyandottes. (D. Scrivener, 2006 – *see* Bibliography.)

Breeding and Management

The Derbyshire Redcap is a very hardy, dual-purpose farmyard fowl ideally suited to low maintenance free-range rearing; it benefits from a large grass run in which to exercise. Derbyshire Redcaps are graceful and well balanced when they are active, and are flighty, often scaling two-metre fences. They can also fly distances of fifteen or twenty metres at a metre or so from the ground if provoked, particularly the bantam version that can be wild in habit.

Eggs are produced in reasonable quantities, are small to medium in size, and white in colour, although tinted eggs also occur, dependent on strain. Breast meat is slow in coming, but is quite substantial for a light breed.

The most distinctive feature of the Derbyshire Redcap is the unusually large and wide rosecomb covered with evenly sized spikes, also known as workings. The comb should be square-fronted, and must not closely overhang or rest on the beak or eyes. Sometimes a specimen may develop a lop-sided comb that covers one eye: this is a fault. There should be a single straight leader level with the top of the comb. The comb should reach 8.25 × 7cm (3.25 × 2.75in) in size, and in shows can account for 25 per cent of the points. The comb may be vulnerable to frost damage during the winter: to avoid damage, ensure that birds have continued access to a dry and insulated house, and if necessary use Vaseline on the combs to offer some protection. Fortunately the rosecomb is much smaller than the original cap of the early twentieth century, which was developed for exhibiting purposes to be 14 × 11.5cm (5.5 × 4.5in) in size. Combs of this size could only hinder the development and comfort of the bird.

The Redcap is very similar in appearance and habit to the Old English Pheasant Fowl, although it is one of very few white egg-laying breeds with red ear lobes: the Old English Pheasant Fowl differs most noticeably in having white ear lobes rather than red, and a smaller rosecomb, and the males have a laced breast, where Redcap males have a solid black breast and underbody. Any white marking or a whitish bloom on the ear lobes of a Derbyshire Redcap is a common fault, and could be a sign of past crossing.

Often Redcap females will develop a pale gingery ground colour, which is not desirable for exhibiting. This may occur naturally when the bird is young, or it may be a result of sun damage. Ideally the ground colour should be a deep nut brown.

As with some other lightweight breeds such as the Hamburgh, Old English Pheasant Fowl and Scots Grey, chicks hatched late in the year from August onwards fare poorly and may not survive the winter.

The bantam Derbyshire Redcap is very alert and quick on its feet; it is particularly rare, and may number less than one hundred (2011). Housing at night and secure fencing would prevent losses to predators, however it has the potential to be a semi-feral variety, being one of the most self-reliant, low appetite, and therefore low maintenance native breeds for those who have the space to let these birds roam free.

The eggs are small to medium in size, and vary from white to near pale brown in colour. Unusually, unlike the large fowl, the bantam is a reliable sitter, capable of rearing her own young.

Main Uses

- **As a dual-purpose breed:** Redcaps will provide a reasonable quantity of medium-sized eggs and the occasional table bird
- **For crossing:** They could be crossed with another more productive breed such as the

Sussex to improve yields whilst retaining a lightweight frame and minimal feed and management requirements

- **For low maintenance, free-range rearing:** They are a hardy breed with a small appetite
- **For exhibition:** The elaborate rosecomb was enhanced for showing in the first instance, and it is still a unique and interesting attribute which may encourage breeders to exhibit their birds

Head of a bantam Derbyshire Redcap pullet, showing how the rosecomb is held high above the beak and eye, with a single straight leader.

Day-old Derbyshire Redcap with correct markings.

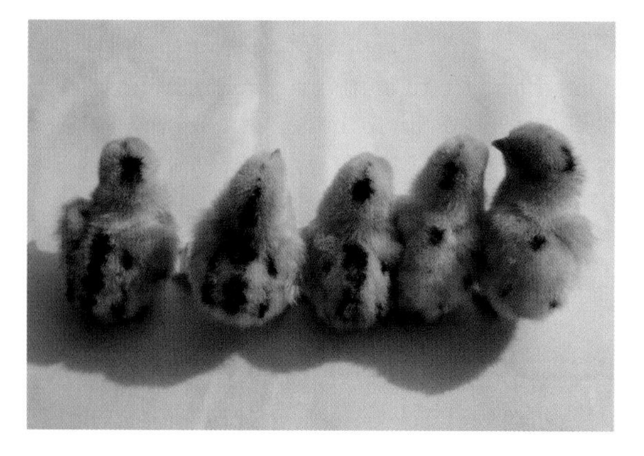

Day-old bantam Derbyshire Redcaps. Some variation can be seen and is an indicator that past crossing has occurred. The bantams still require careful selection to ensure consistency throughout the variety.

A three-week-old cockerel showing how quickly and precisely the comb develops, and how easy it is to identify birds with good combs at a young age.

Rosecomb of a large Derbyshire Redcap cockerel, showing a wide, flat comb with neat and evenly spaced spikes, also known as 'workings'.

A bantam wing: ideally the primaries and secondaries should be red on one half and solid black on the other, with a black tip. In this photo the primaries have black peppering rather than the preferred solid black.

Large fowl Derbyshire Redcap hen with undesirable pale ground colour.

Bantam Derbyshire Redcaps belonging to Christine Taylor of Shropshire.

DORKING

Origin: Ancient British breed
Class: Heavy, soft feather
Colour varieties: Cuckoo, dark, red, silver grey, white
Purpose: Meat and eggs, also historically important
Eggs: Tinted
Weight: Large fowl cock, 4.55–6.35kg (10–14lb); large fowl hen, 3.6–4.55kg (8–10lb)
Ancestry: Not known
Sitter: Yes
Auto-sexing: No. The cuckoo variety cannot be reliably sexed at a day old

Large Red Dorking cock.

History

Year created: Not known, an ancient breed.

Originator: Not known.

Breed development: Birds broadly matching the description of the Dorking with a fifth toe were mentioned as good table fowl by Columella, a Roman writer, when the Romans ruled England at the beginning of the first century AD. The Dorking is one of the oldest British breeds in existence, and along with the Old English Game is likely to have a lineage going back before the time of Jesus Christ. The existing Dorking is likely to be different in many ways to the original farmyard fowl of Julius Caesar's time, in particular as a result of changes made to suit the exhibition scene since the mid-nineteenth century. (D. Scrivener, 2009.)

Bantam: Bantam Dorkings come in all standard colours. They are kept by very few breeders, and most of the strains that currently exist are likely to be relatively recent creations.

Breeding and Management

The Dorking is calm and docile by nature, and best suited to hobbyists who wish to maintain a breed for historical interest, for use as a broody, and in the case of the silver grey variety, as a hardy, general purpose backyard fowl.

The Dorking is traditionally regarded as a table breed and one suited for crossing to create table birds – though unfortunately this is no longer generally the case, particularly with the rarer colours. The silver grey is arguably the only variety that still has the potential to be productive and reliable for the hobbyist – otherwise there are many other native breeds that would be more suitable for table and dual-purpose use, such as the Australorp, Ixworth, Lincolnshire Buff and Sussex.

The Dorking has a moderately long, near-horizontal back and a long, straight keel, giving the impression of a rectangular body and suggesting a type more suited for the table. They are not suited to the hobbyist who wishes to have lots of eggs and an easy time breeding and rearing. Unfortunately poor fertility in cockerels, low resistance to infection, short life

Large Red Dorking hen. Ideally the hen should tend towards red over most of her body. This example is a good type, with a near horizontal back and the tail held close and high.

expectancy and general lack of hardiness are common problems with the rare red, cuckoo and white colours. They can sometimes struggle to gain weight when young, and can lose large amounts of weight when in moult. The rarer colours have been known to die young or randomly, more so than many other breeds, so the value of these rarer colours for crossing is dubious. Undoubtedly the Silver Grey Dorking may still be suitable due to its ongoing popularity and greater vigour, but the other colours can often take two years to fully develop and yet still have little breast meat to offer. It is likely that the ongoing use of these colours for exhibition since the late nineteenth century, and their continued rarity, has had the greatest influence on their reduced productivity and hardiness.

From what the author has seen, egg quality varies greatly in the red variety, both white and tinted eggs can be seen in different stock, and spherical and oval-shaped eggs are common due to the lack of careful selection for egg quality – another indication that in the past, selection was for exhibition purposes only.

The Dorking is one of the few native breeds where different colour varieties display slightly different physical characteristics. The red variety is generally regarded as being slightly smaller than any other, possibly an indication of the size of the original breed before the exhibition era. The silver grey can be seen in traditional and exhibition stock, the latter being often much larger in body with greater fluffiness. Comb type also varies: the dark variety can have a single or rosecomb, the silver grey and red varieties have a single comb, and the cuckoo and white varieties have a rosecomb. The single comb falls over to one side in females.

One of the Dorking's identifying features is that it is one of only two native breeds to have a fifth toe, a condition known as polydactyly (the Lincolnshire Buff being the other). The fifth toe should be close to, but still separate from the fourth toe, and should point upwards. Generally toe placement is good, and this has been achieved by centuries of selection. (The recently recreated Lincolnshire Buff, on the other hand, can have many variations of polydactyly with far less consistency; often the hind toe and fifth toe in the Lincolnshire Buff has webbing or is joined, a possible indication that genetic traits such as the existence of a fifth toe can be fixed and improved, given time.)

The red, dark and white varieties of the Dorking are believed to be the original colours, but the silver grey variety is by far the most popular and supposedly the most productive – all the other colours put together are unlikely to make up the total number of silver greys in existence. The cuckoo is the rarest colour, followed by the white, red and dark varieties, but unfortunately their rarity has led to increased inbreeding and reduced productivity. These colours are now more suited to the devoted breeder who wishes to specialize in a Dorking variety for historical interest.

For those wishing to keep one of the rarer

Dorking colours for posterity, be sure to provide plenty of secure ranging space, well drained or dry soil, and comfortable housing that is kept well insulated or warm over winter. If you are prepared to hatch a lot of eggs and are sure to cull any sub-standard birds then hopefully you may end up with something far hardier – but expect high losses. They are not the most agile or durable of native breeds, so a perch 15–30cm (6–12in) above ground would be adequate.

For the most vigorous youngstock and for the highest survival rates, hatch birds during the spring, as they will then be fully developed before the winter. Birds hatched late in the year, from August onwards, may struggle to grow or even survive during the autumn and winter months.

Main Uses

- **For posterity:** The Dorking is probably the second oldest British breed after the Old English Game. The Dorking may be of historical interest to students, poultry

A breeding pen of Silver Grey Dorkings. (Wernlas collection)

Large Red Dorking cock. This cock may have derived from a cross with the silver grey or dark varieties, the result being the overall lightening in colour. Despite his colour he still breeds correctly marked and coloured chicks.

Day-old Red Dorking chick: these markings are typical of a black-red type chick, similar to the Brown Marsh Daisy and Brown Sussex.

hobbyists and those living locally to Dorking in Surrey

- **For exhibition**
- **As a dual-purpose breed:** The silver grey is the only productive variety; in the other colours, egg-laying ability and egg size is inconsistent. Nor is the breed as useful a table bird as it should be for its size because it is very slow to develop a full breast of meat, and some may not ever make worthwhile table birds despite their size. Generally this is the case with the rarer colour varieties
- **As a sitter (broody)**
- **For crossing:** The silver grey is the most suitable Dorking variety to use for table bird crosses

Fifth toe placement on a Dorking hen. Ideally the ifth toe should be completely separate from the hind toe.

HAMBURGH

Origin: Northern counties of England
Class: Light, soft feather
Colour varieties, British: Silver-spangled, gold-spangled, black, blue
Purpose: Eggs, exhibition
Eggs: White
Weight: Large fowl cock, 2.25kg (5lb); large fowl hen, 1.8kg (4lb)
Ancestry: Not known
Sitter: No
Auto-sexing: No

Large Silver-Spangled Hamburgh male with a good comb, straight tapered leader with a fine point, and near perfect plumage. (Courtesy Robin Creighton)

Head of a bantam Silver-Spangled Hamburgh hen with an undesirable upward-pointing leader to the rosecomb. The leader should be tapered and run level with the top of the comb.

History

Year created: Not known.

Originator: Not known.

Breed development: It is not known why the name 'Hamburgh' was decided upon after the shows of 1848–50, but the name is likely to have derived from the potential close relationship with, or close similarities to similar breeds from Germany and The Netherlands. Before the name 'Hamburgh' was established, the birds were known in northern England by names such as Black Pheasants, Bolton Bays, Bolton Greys and Lancashire Moony Fowls. They were effectively farmyard fowl of similar type but different colours. Breeds such as the Old English Pheasant Fowl and Derbyshire Redcap may well be distantly related to the spangled

Large Black Hamburgh cock.

varieties of Hamburgh, and these breeds were established to amalgamate the various localized forms.

A breed standard was created so that breeders could select their birds to meet a common standard for exhibiting, and to ensure the breeds remained distinct one from another. The pencilled varieties of Hamburgh are effectively a unique Dutch breed that has been adopted within the Hamburgh breed. They are generally slightly smaller than the spangled varieties. The black variety is larger still. The Black Hamburgh was previously known as Black Pheasant Fowl, another variety of energetic farmyard fowl from northern England

Large Black Hamburgh hen, a bird belonging to Charlie Peck of Surrey.

that became amalgamated into the Hamburgh breed and was common around Lancashire and Keighley in West Yorkshire. (D. Scrivener, 2009.)

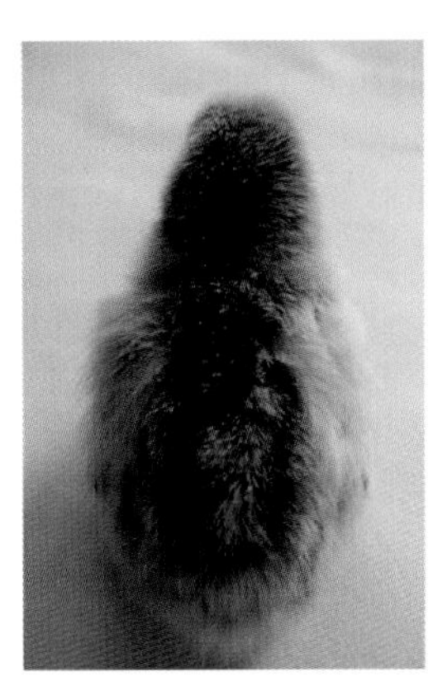

Day-old bantam Silver-Spangled Hamburgh.

Bantam: The bantam Silver-Spangled Hamburgh was initially created by John William Farnsworth of Frampton Place, Boston, Lincolnshire in the late 1880s. He used undersized large Silver-Spangled Hamburgh, Silver Sebright and White Rosecomb. (D. Scrivener, 2009.)

Breeding and Management

The Hamburgh is an energetic and flighty breed suited to low maintenance free-range rearing, kept for its eggs and as an attractive fowl. Hamburghs are quite possibly the most flighty of the native breeds, and can easily scale a two-metre high fence, particularly if they are confined to a small pen. Ideally small pens should be netted over the top to prevent escape. They are an ideal breed for free-range rearing, particularly if they are to be kept for eggs rather than for exhibition.

Exhibition stock ideally requires a small, sheltered pen, perhaps with solid sides to prevent damage to the ear lobes which would otherwise occur if they were confined in a wire or weldmesh run. A small sheltered pen would also help to ensure the plumage remains in prime condition. Silver-Spangled Hamburghs may suffer from some brassiness to the silver-white ground colour if they are left out in the sun. Hamburghs are an active breed so running space and freedom to forage would suit their habit.

The faulty tail of a cock bird showing black ticking and smuttiness on the white of the feather before the black spangles.

The Black Hamburgh is derived from the Black Pheasant Fowl of Yorkshire and is the least common of the Hamburgh varieties. The diminishing numbers of Black Hamburgh is likely to have resulted in the spangled varieties being bred into the black to keep the variety going. The silver-spangled variety is the most common, but both silver and gold varieties are commonly exhibited. Both large fowl and bantam versions exist, the bantam spangled varieties being more common than the large fowl. There is much variation in bantam size. Many bantam Hamburghs are above the weight limit as described by the standard, but not large enough to be classed as a large fowl. It is then important for breeders to select by size and weight to maintain the correct bantam size.

Hamburghs should have a smooth face free from hairs. Earlobes should be smooth and flat, and vary in size depending on colour variety. In exhibition birds, lobes should be pure white without any scabbing or sign of red. Black Hamburghs have the largest lobes, possibly accentuated by past crossings with Minorcas. The legs are lead blue, although black legs are acceptable in the black variety. Black Hamburghs have an all-over beetle-green sheen: there should be no purple. A white throat – also known as 'bishop throat' – on the silver-spangled variety is a rarely seen fault.

The male and female spangled varieties have large, heavy spangles at the tips of each feather. The spangles are larger than those of the Old English Pheasant Fowl or Derbyshire Redcap.

Being a small lightweight breed, chicks should be reared separately from larger breeds such as the Sussex to prevent unfair competition and to ensure healthy growth. Chicks do not fare well when hatched late in the year, from August onwards.

Main Uses

- **As a productive egg layer:** Hamburghs have been known to be reasonable layers, but this will depend on the strain. Large fowl eggs are quite small, and the bantam eggs are even smaller, so for utility value it may be more useful to keep the large fowl
- **For crossing:** Crossed with a more productive laying breed such as the Sussex or Ixworth could yield more productive layers, and with sensible selection, a small lightweight hybrid could be created with reduced appetite
- **For exhibition:** The Hamburgh has been predominantly kept and selected for exhibiting, and requires attention to detail to ensure clean, smut-free plumage with accurate and consistent spangle markings
- **For low maintenance, free-range rearing**

Bantam Silver-Spangled Hamburgh cockerel. The comb has a good square front, level top, plenty of small rounded spikes of even height, and a tapered leader level with the top of the comb. The plumage is acceptable, but there is too much black in the tail coverts.

INDIAN GAME

Origin: Cornwall, England
Class: Heavy, hard feather
Colour varieties: Dark, jubilee, double-laced blue
Purpose: Meat
Eggs: Tinted
Weight: Large fowl cock, 3.6kg (8lb) min.; large fowl hen, 2.7kg (6lb) min.
Ancestry: Asil, Malay, Old English Game (Pit Game)
Sitter: Yes
Auto-sexing: No

Large Dark Indian Game cockerel.

History

Year created: The Indian Game was officially recognized in 1886.

Originator: Dark variety not known.

Large Dark Indian Game hen.

Breed development: The Indian Game was developed in Cornwall early in the nineteenth century by crossing Old English Game, Asil and Malay. The Jubilee Indian Game was developed by Henry Hunt of Springfield Farm, Clifford Meane, Gloucestershire; his work started in 1892, and the variety was first shown in 1897 to commemorate Queen Victoria's Diamond Jubilee.

The Indian Game was developed for exhibition and the table; despite its relatively slow growth rate it cannot be matched by any other traditional British pure breed for sheer quantity of breast meat. The wide and prominent breast and thick body are obvious characteristics of the breed, but in the early years of their existence they were somewhat taller and longer in the shank and thigh. The cockerels were traditionally used in crossing with other native table breeds such as the Sussex and Dorking that are lighter boned and also have white skin and shanks, more desirable for the UK market. (D. Scrivener, 2009.)

Bantam: The bantam Indian Game was created by W. F. Entwistle of Wakefield, Yorkshire and first shown in 1887. (D. Scrivener, 2009.)

Breeding and Management

The Indian Game is a durable breed with a slightly nervous and lively disposition; they are used as table birds, and for crossing to create faster growing meat birds. They are generally docile and not as aggressive as the Game breeds used in their development. Indians, known as Cornish in the USA, are persistent sitters and defensive broodies.

Indian Game are wide across the shoulder with the body tapering smartly to the tail. They are firm fleshed and strong. The legs and feet must be steady and strong, the feet should be well spaced, and the bird must stand and walk well. They are successful free range birds and will benefit from having a large pen to exercise. The breed type and size must not be exaggerated to a point that it can no longer reproduce naturally. A tall, long-legged male should improve fertility as treading is made easier.

Indian Game chicks do not feather up as quickly as most other native breeds but they do develop a bulky and heavy body from early on. Often chicks come in different colours, depending on the source of stock, and this is an obvious sign of past outcrossing, either from sex-linked crosses or from crossing with other Asian Game breeds. All chicks should have a smoky yellow down with indistinct dark brownish markings along the back.

The body type is dominant, meaning that the offspring from an Indian Game cross should have the desirable wide, meat-producing body conformation of the Indian Game.

The breed is prone to broodiness, and they are reliable, tight sitters but are clumsier with eggs than other breeds.

The Indian Game comes in three standardized colours: dark, jubilee and double-laced blue. The dark variety is by far the most common. Jubilee and blue-laced are rarely seen but still exist, particularly as exhibition stock. To maintain the quality of the lacing on the jubilee variety, it has been traditional to cross with the Dark Indian Game; this crossing has ultimately resulted in many Dark Indian Game birds

The head of a Dark Indian Game male with a pea comb and red face, lobes and wattles.

Large Dark Indian Game cockerel, broad across the shoulders, with tight feathering and a wonderful glossy green sheen.

Large Dark Indian Game hen.

Large Jubilee Indian Game hen. (Wernlas collection)

having occasional white feathers, sometimes in the wing or on the neck.

Currently, many birds being sold as Indian Game have characteristics that are not true of the breed. White legs are commonly seen and should be avoided as it may indicate the history of a crossing with the Sussex or another breed. Ideally the legs should be rich orange or yellow. Leg colour may become paler late in the year, as the yellow pigment carotene is used extensively in the yolk of eggs.

Reddening of the skin along the breast bone is common due to the lack of feathering along the keel. Reddening is a common characteristic of the breed and not a hindrance, but it is unappealing on a plucked carcass. When plucked, the Dark Indian Game will leave some undesirable (in traditional thinking) dark stubs.

The Indian Game has a pea comb: this is a comb with three longitudinal ridges, the middle one being twice the height of the other two. The comb should be small and held close to the skull.

Main Uses

- **As a table bird:** The Indian Game develops a full, broad breast of meat. Its meaty and cobby type is noticeable when it is handled, even at two or three weeks of age
- **As a sitter (broody)**

A day-old Dark Indian Game chick with what is believed to be the correct down colour, with indistinct faint dark markings on the head and along the back.

ABOVE: White feathers in the neck hackle are a common fault in male and female stock, possibly due to past crossing with the jubilee variety.

OPPOSITE: A Dark Indian Game hen with excessively wide, black laced markings on her back.

The breast of a large Dark Indian Game cockerel with sparse feather coverage along the breastbone and reddened skin.

ABOVE: Faulty Dark Indian Game hen, with undesirable pale ground colour and white legs.

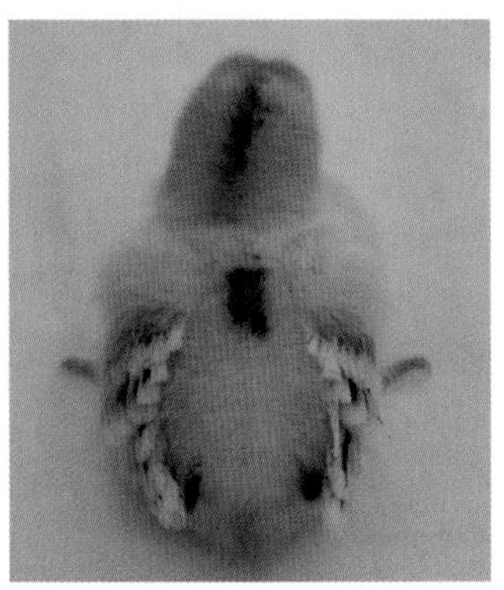

Another common down colour of the Dark Indian Game. The head and tail are a darker shade to the rest of the body.

THE IXWORTH

Origin: Village of Ixworth, Suffolk, England
Class: Heavy, soft feather
Colour varieties: White
Purpose: Eggs, meat
Eggs: Tinted
Weight: Large fowl cock, 4.1kg (9lb); cockerel, 3.6kg (8lb); large fowl hen, 3.2kg (7lb); pullet, 2.7kg (6lb)
Ancestry: Indian Game, White Sussex, White Old English Game, White Orpington, White Minorca
Sitter: Yes
Auto-sexing: No

History

Year created: Standardized in 1939.

Originator: Reginald Appleyard.

Breed development: The Ixworth was developed as a specialist table bird. By using white breeds to create a breed with white plumage, Reginald Appleyard was able to ensure a clean-looking carcass once plucked. The Indian Game and Sussex gave the Ixworth a broad, full breast of meat, and the use of the Minorca ensured that the breed remained active and energetic. The Ixworth also has an excellent egg-laying ability derived from the Sussex, Minorca and Orpington, and is currently one of the most productive dual-purpose breeds.

The breeds chosen by Reginald Appleyard were carefully selected and crossed in an attempt in the pre-broiler era to develop a bird specifically for meat production, with desirable traits such as white skin and white plumage. However, the development of productive broilers in the USA happened more quickly than in the UK, due to World War II. The American broilers were derived from the Indian Game (Cornish) and the Plymouth Rock, and were carefully selected and quickly improved for faster growth rates and better feed conversion ratios. (D. Scrivener, 2006.)

The Ixworth never truly got off the ground and never became a serious commercial concern. If it had been developed as a pure breed twenty years earlier, it would have been a huge success.

Bantam: A bantam Ixworth was believed to have been first exhibited by Reginald Appleyard in 1949. Unfortunately being a relatively plain white bird, it never became popular, and probably became extinct soon after. (D. Scrivener, 2006.)

Large Ixworth cock. *Large Ixworth hen.* *Large Ixworth cockerel.*

Breeding and Management

The Ixworth's name is pronounced 'Ixs'worth as opposed to 'Ick'worth. The breed has excellent egg-laying abilities due to its Sussex and Minorca ancestry, often producing eggs with a slight rosy tint; they can be very large, but this depends on the strain. The Ixworth is a white-skinned table bird that produces reasonable quantities of breast meat, helped by its wide breast derived from its Indian Game parentage.

The Ixworth is attractive, with a proud posture and a firm, cobby body; it has a lively and hardy ranging habit, and has a hearty appetite. It is not aggressive or flighty. It comprises a good amalgam of characteristics and habits from each of its parent breeds. After many years of being close to extinction, the breed has now become fashionable, primarily due to its value as an excellent dual-purpose bird for the smallholder. It would seem that the main reason for its past rarity was its plain white colour, many people at first glance regarding them as just another modern broiler.

Currently there does appear to be some variation in type, size and quantity of breast meat among existing stock, and this is most likely due to the outcrossing of some strains with other white breeds or hybrids. Any sign of yellow in the legs would be a clear indication of past crossing, probably with a modern broiler. A young specimen with an unusually large amount of breast meat for its age, by traditional poultry standards, may also suggest recent crossing.

The Ixworth has a pea comb, the size of which varies depending on the specimen; in some it may fall to one side, which is a fault. A small comb held close to the skull would be more suitable and would follow that of the Indian Game. The pea comb is made up of three longitudinal ridges, the middle one being the most prominent.

The legs are white and often have pink-red shading along the shank and on the underside of the feet. Plumage is compact and semi-tight; there should not be any sign of fluffiness. Birds kept at free range in full sun may develop some yellowing – also known as brassiness – to the plumage. The white plumage allows the bird to be plucked clean, without leaving undesirable dark stubs.

The Ixworth does not require special housing requirements, and can be kept in an ark or on a secure open field. It benefits from plenty of exercise to keep it fit, as it can be a heavy bird, and although leg and heart problems are rare, they are more likely to occur with the Ixworth than many other native breeds.

Main Uses

- **As a dual-purpose breed:** The Ixworth is potentially the most useful native breed for the domestic user
- **As a sitter (broody):** It is a persistent sitter
- **For free-range rearing**

Pea comb of an Ixworth cock, also known as the triple comb. Three ideally straight longitudinal ridges make up the comb, the middle one being the most prominent.

LINCOLNSHIRE BUFF

Origin: Lincolnshire, England
Class: Heavy, soft feather
Colour varieties: Buff
Purpose: Eggs, meat
Eggs: Tinted
Weight: Large fowl cock, 4–5kg (9–11lb); cockerel, 3.1–4kg (7–9lb); large fowl hen, 3.1–4kg (7–9lb); pullet, 2.9–3.6kg (6–8lb)
Ancestry: Buff Cochin, Red Dorking
Sitter: Yes
Auto-sexing: No

Lincolnshire Buff pullet.

History

Year created: The breed was created in the second half of the nineteenth century; it was standardized in 1997.

Originator: Unknown. The recreated breed was primarily developed and promoted by Brian Sands of Boston, Lincolnshire.

Breed development: It is thought the Lincolnshire Buff was created in the 1850s on Lincolnshire farms and smallholdings from the crossing of the local farmyard breeds and the original Buff Cochin, the Buff Shanghai. The breed was developed as a table bird for the London market, being transported by train. Unfortunately the original breed was never standardized and eventually became extinct.

Large Lincolnshire Buff cockerel.

In 1981 the Small Animals Unit at the Lincolnshire College of Agriculture and Horticulture and Riseholme Agricultural College started a project to recreate the Lincolnshire Buff. When the programme closed in 1986 Mr Brian Sands of Lincolnshire, along with other keen breeders, continued this work. Brian recovered one cockerel and four hens from Riseholme, and later introduced Buff Cochin and Red Dorking to finalize the recreation. A breed society was established in 1995. (D. Scrivener, 2006.)

Bantam: None exists.

Breeding and Management

The Lincolnshire Buff is a very large and docile dual-purpose breed. It lays large eggs that are often tinted or pale brown (officially they should be tinted – off-white). It is one of the largest native breeds, but careful selection is required to increase the meat quantities. The buff plumage is accompanied by umber in the tail and main wing feathers and axials; it is never profuse and plucks out clean, making for a presentable carcass. The breed can be distinguished from other buff breeds by its fifth toe and its large semi-tight feathered body.

The hens make very good brooding mothers due to their large size and brooding instinct.

Despite the creation of a society the Lincolnshire Buff has become very rare, with very few people breeding them. The lack of interest is unfortunate, because the breed is capable of being perhaps one of our best native all-round performers for domestic use. Their calm nature makes them ideally suited to being kept with children, and they can be as easy to handle as the Buff Orpington.

Fifth toe placement is an important factor in selecting, stabilizing and exhibiting the breed. Polydactyly is the trait that causes the hind toe to duplicate, but the expression of polydactyly is irregular and may result in duplication on only one foot; or sometimes only the claw may be duplicated. This sort of variation is more prevalent in the Lincolnshire Buff than in the Dorking. It may be due to the Dorking's great antiquity that greater consistency in fifth toe placement can be seen, and it may take many years of careful selection for the same consistency to be seen in the Lincolnshire Buff.

Black tipping forming a fringe around the base of the neck is a fault, and could be a sign of past crossing with another useful utility breed, the New Hampshire Red.

Lincolnshire Buff cockerel.

The wing of a large Lincolnshire Buff cockerel; some umber is acceptable on the main wing feathers.

The fertility of cockerels is often short-lived, and only sporadic after two years of age. This is a problem associated with several native breeds that are either excessively inbred or of poor consistency. The Dorking, Marsh Daisy and Sebright are other breeds that may have reoccurring fertility problems in the population, a subject that may cause some

Bleached and patchy plumage caused by exposure to full sun while running free range.

Good fifth toe placement. The fifth toe is straight, points upwards, and there is no webbing.

The fifth toe is partially joined to the hind toe by webbing. Placement like this is common and not desirable on birds intended for exhibition.

Undesirable placement. The fifth toe points down.

This bird has a partially formed sixth toe on both feet. A sixth toe is rare and not suitable in breeding stock.

debate. The gene pool for Lincolnshire Buffs is very small as they are a recently recreated breed developed with limited stock, so it is not necessary or perhaps possible to source completely different lines. However, the New Hampshire Red is likely to have been used to improve colour and increase the gene pool in some strains of the Lincolnshire Buff. The great variation in buff shades that can sometimes be seen may be testament to this, along with occasional dark tipping to the lower neck hackle. If crossing with the New Hampshire Red has occurred it would only aid the utility merit of the Lincolnshire Buff.

Buff plumage can become damaged by the sun, resulting in mealiness. Birds under one year of age have the most consistent buff colour and are best suited to exhibiting. If plumage colour is of importance to the hobbyist, it will be necessary to provide plenty of shade in a pen, perhaps in the form of established trees. When selecting breeding stock, ensure that both male and females have a similar shade of buff, because dark buff specimens crossed with pale buff will result in patchy and mealy-coloured offspring.

Main Uses

- **As a dual-purpose breed:** The Lincolnshire Buff is a more prolific producer of large eggs than it is of meat, but it is still one of the most useful all-round native breeds. It has a greater utility merit than the Orpington
- **As a sitter (broody):** The hen is a persistent sitter capable of covering a clutch of at least ten eggs

MARSH DAISY

Origin: Lancashire, England
Class: Light, soft feather
Colour varieties: Black, brown, buff, wheaten, white
Purpose: Eggs
Eggs: Tinted
Weight: Large fowl cock, 2.5–2.95kg (5.5–6.5lb); large fowl hen, 2–2.5kg (4.5–5.5lb)
Ancestry: White Leghorn, Black Hamburgh, Old English Game Bantam, Malay, Old English Game (Pit Game), Sicilian Buttercup
Sitter: Yes
Auto-sexing: No

Wheaten Marsh Daisy cockerel.

History

Year created: The Marsh Daisy was first standardized in December 1922.

Originator: Mr John Wright of Marshside, Southport, Lancashire and Mr Charles Moore of Hatfield Woodhouse, Doncaster.

Breed development: The long process to create the breed was unwittingly started by John Wright in 1880. John Wright maintained a small flock of Old English Game bantams which he crossed with a Cinnamon Malay cock. A cockerel produced from that cross was given to a Mr Wignell, who mated the cockerel to his White Leghorn hens. A cockerel derived from this cross was given to John Wright and later crossed with a flock of Rosecombed Leghorns which came from a previous crossing between a Black Hamburgh cock and White Leghorn hens.

A closed flock of the offspring was maintained for thirty years.

In September 1913 Mr Charles Moore purchased two hens from John Wright and mated them with a Pit Game cock as there were no cocks of the white strain remaining in John Wright's flock. The offspring were mated with a Sicilian Buttercup, which fixed the green legs, and this led to the creation of the Marsh Daisy.

Wheaten Marsh Daisy hen.

The Sicilian Buttercup was most probably used because it was an exciting new breed, introduced into the UK by a Mrs Colbeck in 1912. The wheaten, white and buff were the first colours, and were created by Charles Moore of Hatfield Woodhouse near

Doncaster at about the same time. The brown was developed in the mid-1920s by a Lord Canterbury who lived in Norfolk (not Kent as the name may suggest). The black was probably created by Mr E. C. Parsons of Williton in Somerset at about the same time as the brown. (D. Scrivener, 2006.)

Bantam: None exists.

Breeding and Management

The Marsh Daisy is a very friendly and docile breed, neither flighty nor aggressive. The hens can be used purely for back-garden egg production, in which capacity they will continue to be productive over many years, although not usually in such great volume as once was recognized for the breed. Unfortunately egg size is small, and egg colour varies greatly depending on the strain, from pure white to tinted and pale brown.

The main concern with the Marsh Daisy is its inability to breed true, and as a result it is one of few native breeds that is on the downward slope to extinction, and has been for many years. Although the Marsh Daisy comes in many colours, they were probably never truly fixed in the first place, the result of an overwhelming desire at the time to get the breed known to the public, to promote its qualities and attempt to make it commercially viable as quickly as possible. In hindsight, a common mistake of many newly developed breeds or varieties of the early twentieth century is that they were released before they were fixed and breeding true; for example the Norfolk Grey had breast lacing on some specimens shortly after being launched to the public.

The problem today is that the existing colours have been bred together, in particular the wheaten and brown varieties; therefore for dedicated breeders many years of selection lie ahead to ensure that each colour breeds true. A mating of a male and female that exactly match the breed standard will still yield a mixed bunch of chicks, with wheaten and brown colours combined on one bird, and some with random black markings; occasionally whole buff and pile birds will also appear amongst the offspring of a single clutch. Pile-coloured specimens that have red pigment bleeding through the white on the breast of females and the shoulder and back of males are recessive whites. Single combs and five-toed specimens have also been found among some offspring, a clear indication of past cross-breeding.

It would appear that the existing stock is widespread, though with denser pockets of birds around Norfolk and Yorkshire. Unfortunately, due to the forgotten state of the mysterious Marsh Daisy, much selection is still required to refine the breed and its features to the way they should be. It is important to recognize that Marsh Daisies, even those that have been carefully sourced and selected, could still throw a multitude of colour variations.

The Marsh Daisy is therefore a breed more suited to enthusiasts who like a challenge and wish to be influential in the recovery and promotion of a particularly rare British breed. They must be prepared to hatch large numbers each year, and to dispose of all those that do not match the requirements of the breed standard, preferably by culling – and that will be a large proportion. The birds photographed almost meet the required standard, but still have a little room for improvement. Red plumage, for example, is a common fault, often seen as a darkening of the rich or dark gold of the wing bow on males.

It would appear that the brown variety is the most common. Typically wheaten varieties of other breeds have black-red males – males with a black breast and underbody – whereas Wheaten Marsh Daisy cockerels have a golden-brown breast and underbody. So it is not hard

Brown Marsh Daisy cock with a correctly sized comb neatly set high on the head with a 13mm (½in) leader level with the surface but not straight. Unfortunately the workings appear more like folds.

Buff Marsh Daisy cockerel, a bird bred by Clive Greatorex of Cambridgeshire.

Buff Marsh Daisy pullet, as near to a consistent pure buff that can currently be found. Bronze or black tail feathers in both sexes are common at present.

Brown Marsh Daisy hen. Ideally the dark peppered colour of the back feathers should extend more, over the wing bow.

ABOVE LEFT: *Brown Marsh Daisy chick.* ABOVE RIGHT: *Day-old Wheaten Marsh Daisy. Chicks like this one are typically coloured for a wheaten chick, yet the resulting bird may develop some plumage markings similar to one of the other varieties, such as brown.*

to see that some breeders may have been breeding brown cockerels with wheaten hens, Brown Marsh Daisy males also being of the black-red type.

The breed standard requires willow green legs, although many shades of green, grey and yellow are seen. Often birds of the buff variety will have pale green or yellow legs, whereas birds of the brown variety often have dark green or grey legs. The crossing of black- or grey-legged birds with yellow-legged ones will yield green-legged birds. It would be too easy to get overwhelmed with the shade of green required for Marsh

Pile Marsh Daisy pullet. Sometimes these recessive whites are thrown from the general Marsh Daisy population.

Wheaten Marsh Daisy hen.

A wheaten hen showing the back feathers, with white wheat edging with a red wheat centre and black tipping to the chestnut hackle, forming a fringe at the base of the neck.

A poorly coloured buff bred from buffs. The type is good but the colour is inconsistent, and features some pencilling to the back and saddle feathers and black tipping on the neck hackle.

An example of the sort of random mixed colours that frequently occur, even when standard quality specimens of the same colour are bred together.

Brown Marsh Daisy hen showing blue-grey undercolour and brown back and wing feathers with dark brown or black peppering. Ideally a brown hen would have peppered feathers like these on the back and wings, however at present the wing feathers are usually redder.

After the first adult moult Buff Marsh Daisy cockerels sometimes currently develop white feathering through the buff, predominantly on the shoulder, wingbow and back.

A common but very poor type of rosecomb found in Marsh Daisies at present, described by Mr Charlie Peck as being a 'slug'. The comb has no desirable workings (spikes) or a single straight leader, and is not quite wide enough.

A good-sized comb, but needing more workings. The leader is nearly split in three which may be in some way linked to the cup comb and past reintroductions of the Sicilian Buttercup.

This comb of a Brown Marsh Daisy owned by Charlie Peck is very near to the sort required of the breed standard, but more spikes are needed on the surface. A comb with workings similar to that of the Old English Pheasant Fowl would be ideal, but is rarely seen.

Daisies, and since the breed standard does not specify the required shade, it is only fair to assume that breeders can select their own preferred shade of willow green.

Ear lobes should be pure white, smooth and almond-shaped. Less than a third white in the ear lobe is a fault.

Buff plumage is easily damaged by the sun, so Buff Marsh Daisies would benefit from being kept in a shaded run if they are to look their best.

The Marsh Daisy was developed to be a utility breed, and this is clear by the breed standard's inclusion of twenty points for laying ability; therefore when a bird is exhibited, its potential to be a good layer is also assessed. Unfortunately the breed was not productive enough to challenge the egg-laying ability of well established breeds such as the Sussex, and so gradually fell out of favour.

Today if the breed were to be stabilized the Marsh Daisy could prove exceptionally useful as a low appetite breed, a persistent layer of eggs over many years, and as a friendly breed for children.

Main Uses

- **As a sitter (broody):** The Marsh Daisy is a persistent sitter capable of covering eight medium-sized eggs comfortably
- **As a productive egg layer:** Consistent production over many years is still a common feature of the breed
- **For preservation and improvement:** The breed is a challenge for keen poultry keepers
- **For low maintenance, free-range rearing**

MODERN GAME

Origin: United Kingdom
Class: Hard feather
Colour varieties: Birchen, black, black-red, blue, blue-red, brown-red, gold duckwing, silver duckwing, lemon-blue, pile, silver-blue, wheaten, white
Purpose: Exhibition
Eggs: Tinted
Weight: Large fowl cock, 3.2–4.1kg (7–9lb); large fowl hen, 2.75–3.2kg (5–7lb)
Ancestry: Old English Game (Pit Game), Malay
Sitter: Yes
Auto-sexing: No

Bantam black-red Modern Game male. (Courtesy Simon Mckean)

History

Year created: Developed gradually from 1849 when cock fighting was banned, to the beginning of the twentieth century.

Originator: Not one person, but many breeders and exhibitors of Old English Game were responsible for its development.

Breed development: Cock-fighting breeders turned their attention to exhibitions once cock

Bantam black-red (Partridge bred) female. (Courtesy Simon Mckean)

fighting was banned, and a taller, tighter-feathered game breed gradually evolved. The Malay is likely to have been added to enhance the upright carriage of the Old English Game to compete on the show bench, eventually leading to the creation of a new breed, the Modern Game.

Bantam: It is uncertain who were the original creators of the bantam variety, but bantams now come in all standardized colours.
Bantam weights: Males, 570–620g (20–22oz); females, 450–510g (16–18oz).

Breeding and Management

The Modern Game has been developed purely for exhibiting as a fancy fowl. They are a calm-natured breed and are not destructive or aggressive. They have a long, slightly arched neck. The body is wide across the shoulders, tapering to a short tail, with a short, flat back. Birds reared in confined areas can develop a fatty rather than a flat back, so exercise and a controlled diet is essential.

A well balanced conformation is also essential. The legs are long and should have a slight bend at the hock, rather than being dead straight.

The hind (fourth) toe should rest flat on the ground and must not be held close to the foot (duck-footed). Leg colour varies depending on the colour variety, and must be correct for the variety.

The breed standard gives precise details as to the appearance of the breed and its many varieties. The comb is of the single type in all varieties. The comb, ear lobe, face and wattle colour vary, depending on the colour variety of the breed; red and dark purple or mulberry, also known as 'gypsy-faced', are common. Currently most birds exhibited are following a fashion for a finer-bodied bird, so are slightly lighter than the breed standard demands.

Good management and regular handling are essential to ensure that birds remain calm and stand correctly when on display.

Colour is an important factor when birds are judged. To avoid incorrect markings and general wear and tear it is usual practice to exhibit birds that are less than one year old. Continued exposure to direct sunlight will damage plumage colour, so plenty of shade or shelter is helpful. A balanced and nutritional diet will also ensure their plumage remains in top condition.

The tight-feathered plumage of the Modern Game means they are often less susceptible to lice than many soft-feathered breeds. They are also a sitter, but their close feathering limits the number of eggs that can be comfortably held beneath a female when she is brooding.

To keep Modern Game in prime condition they should be kept in a confined environment such as a shed or aviary with plenty of head room for them to stand naturally. To prevent leg weaknesses, which may occur occasionally, ensure birds have enough room to exercise: a small cage is not ideal for long-term management. Birds reared in cages often develop large deposits of fat near the vent, so exercise is essential for birds to remain fit and healthy.

Unfortunately their long legs, conformation and the continued effects of inbreeding and breeding for exhibition means they do not make a durable free-range breed for exposed sites; however, they will cope in a sheltered garden, for example.

Egg-laying ability varies depending on the strain and even the colour variety kept. Egg

THE PRACTICE OF 'DUBBING'

Dubbing involves the removal of the wattles, comb and ear lobes of male Modern Game and Old English Game, including any blemishes. Traditionally dubbing is carried out when a cockerel's sickles are equal in length to the main tail feathers, when it is about six months old. A pair of curved scissors is suitable to carry out the procedure. The comb is cut back close to the top of the head to create a clean-looking, snake-like face.

Dubbing has traditionally taken place since the time of cock fighting, and originally prevented the comb, wattles and ear lobes becoming damaged during a fight. Nowadays dubbing is a controversial subject. It is still carried out by exhibitors, but since the hobbyists who specialize in rearing Modern Game and Old English Game always keep mature males separate to prevent fighting, in fact it serves no practical use. The comb and wattles help regulate the bird's body temperature, so their removal puts it at a significant disadvantage during hot weather. The Animal Welfare Act of 2007 states that keepers of Modern Game and Old English Game may continue to dub their birds providing it is done before they are seventy-two hours old, but after this time only a veterinary surgeon is permitted to carry out the procedure, to ensure the birds are dubbed safely and as painlessly as possible, and to prevent infection.

colour can also vary from white to various tinted shades, but officially eggs should be tinted.

Modern Game should be subjected to the practice of 'dubbing' if they are to be exhibited: currently, as of 2011, the practice is still legal in Great Britain.

Main Use

Exhibition: The Modern Game has been specifically developed for showing

Bantam Pile male. (Courtesy Simon Mckean)

Bantam Silver Duckwing male: the legs in the picture are a little too straight; ideally there should be a slight bend to convey the impression of readiness and activity. (Courtesy Simon Mckean)

Pile Modern Game chicks of varying shades.

Black-red Modern Game chick.

A dubbed head of a Black-Red male showing that the comb and wattles have been removed.

NORFOLK GREY

Origin: Norwich, Norfolk, England
Class: Heavy, soft feather
Colour varieties: Black-breasted, birchen grey
Purpose: Eggs, meat
Eggs: Tinted or pale brown
Weight: Large fowl cock, 3.2–3.6kg (7–8lb); large fowl hen, 2.25–2.7kg (5–6lb)
Ancestry: Birchen Old English Game, Partridge Wyandotte, and possibly other unrecorded breeds
Sitter: Yes
Auto-sexing: No

Norfolk Grey hen.

Large Norfolk Grey cock: this bird comes from the original strain kept by the Reverend Andrew Bowden.

History

Year created: The Norfolk Grey was developed before 1914, before World War I. It was first shown at the 1920 dairy show.

Originator: Mr Frederick Myhill from Hethel, near Norwich.

Breed development: The Norfolk Grey was originally called the Black Maria (pronounced 'Mar-e-a'), the nickname for the German shells used in the trenches of World War I, due to the black smoke from the explosion (Mr Myhill having served in this war). However, there are other possibilities for the name choice; in any case the name Black Maria did not catch on so the name was changed to Norfolk Grey in about 1925.

The Norfolk Grey was developed as a utility breed for the table and to a lesser extent for eggs, but it never became a popular commercial breed and slowly went into decline until they were believed by many to be extinct. Fortunately Susan Bowden rediscovered them at Firs Farm, Farnborough, Banbury, Oxfordshire in 1973. The Rev. Andrew Bowden decided to buy the last remaining cock and three hens for £8. To reinvigorate the strain they saved, Andrew Bowden decided to introduce an Australorp hen.

Since the 1970s there has been much controversy as to whether the Norfolk Grey has a constant lineage dating back to the original stock, or if it is merely a recreated breed. Fortunately due to the work of the Bowdens and recently Frank Bridgland of Amlwch, on the Isle of Anglesey – a man determined to find out the true history of the breed – the Norfolk Grey is now rightly recognized as a rare breed, probably descended from the original stock.

There remains some variation in stock, perhaps due to more recent cross-breeding.

Stock can vary in size, suggesting that some lines, particularly of larger stock, may be recreated, perhaps using breeds such as the Australorp or the Silver Sussex.

Bantam: None currently exists.

Norfolk Grey cock. A large bird, which superficially looks like a Norfolk Grey but has some impurity. This cock used on pure Norfolk Grey hens has yielded gold hackled specimens.

Breeding and Management

The Norfolk Grey has the potential to be one of our most useful and hardy native breeds for domestic use. Classed as a heavy breed, they are in fact one of the lightest of the heavies, and are a prolific producer of medium to large eggs. Egg colour varies from tinted to a milky coffee colour, depending on the strain, but ideally eggs should be tinted. It is likely that eggs coloured pale brown are the result of the past crossing with the Australorp.

Norfolk Greys do not carry enough meat by today's standards to be worth using as a table bird; furthermore black feathering plucks out to leave dark stubs, which some people traditionally consider to be unsightly. Nonetheless, with careful selection or by crossing, a highly versatile dual-purpose breed could be created.

The energetic foraging nature of Norfolk Greys is ideally suited to low maintenance, free-range rearing, much like the Derbyshire Redcap or Old English Pheasant Fowl, but they are more suited to hobbyists with smaller gardens as they are less flighty.

The white feathering on the Norfolk Grey can become yellowed by the sun: this is known as 'brassiness' and is a fault, so if birds are to be exhibited, it is useful to have a shaded run.

A hen and pullet. The hen on the left is the better of the two birds because she has the correct dark eye colour and a more precisely marked neck hackle.

Day-old Norfolk Grey chick.

Head of Norfolk Grey cock showing the desirable dark eye.

A cherry blossom tint can sometimes be seen on the feet and shanks. Legs are slate-black in colour which is acceptable, although pure black is preferred.

The males have more white plumage than the females so are particularly susceptible to brassiness, which usually starts to show in the second year.

The degree of silver-white edging to the neck hackle on females can vary, and often does not continue to the tip of the hackle feathers. The few exhibition strains that do exist have more consistent and more accurately marked female offspring, but the stock is also larger framed.

The origin and purity of some Norfolk Grey stock is dubious. If incorrectly coloured stock is produced, such as whole white birds or a gold rather than a silver neck hackle, then the stock is impure. Such anomalies can sometimes be derived from stock that exactly meets the breed standard.

Main Uses

- **As a dual-purpose breed**
- **As a sitter (broody):** The Norfolk Grey is less likely to go broody than the Indian Game, Orpington or Lincolnshire Buff, but is still useful as such
- **For low maintenance, free-range rearing:** A heavy breed, but more energetic than the Sussex, Ixworth, Orpington and others
- **When crossed:** The Norfolk Grey may prove a useful breed when crossed with an Indian Game sire for table use, perhaps retaining a 'heavy weight' while remaining smaller than breeds such as the Sussex, but with increased breast size. Currently the breast meat is not abundant enough to be worthwhile for most hobbyists

RIGHT: *Breast lacing and shaftiness: this is a serious fault, and one that was common in the early years of the Norfolk Grey's development before 1930. Perhaps this is a clear sign that the Wyandotte was used in their creation.*

BOTTOM LEFT: *Norfolk Grey pullet, with a gold hackle. Knowing some of the history of the cock bird that produced this pullet, it is possible that the gold has come from accidental Marsh Daisy influence several generations back.*

BOTTOM RIGHT: *Norfolk White: it is likely that the white has been thrown – on several occasions in the instance of one breeder's flock – by past crossing with another breed. The only other breed kept by the original breeder was the Marsh Daisy, a breed with a very complex background.*

OLD ENGLISH GAME BANTAM

Origin: England
Class: Hard feather
Colour varieties: Twenty-three
Purpose: Exhibition
Eggs: Tinted
Weight: Cock, 620–740g (22–26oz); hen, 510–620g (18–22oz)
Ancestry: Old English Game Oxford/Carlisle
Sitter: Yes
Auto-sexing: No

Spangle Old English Game bantam cockerel.

History

Year created: Not known. The Old English Game Bantam was first exhibited under the name in the mid to late 1890s.

Originator: Not known.

Breed development: Bantam Game fowl have existed throughout Europe for centuries and

Spangle Old English Game bantam pullet: the carriage is good but the colour requires more white tipping to the breast feathers.

have gradually evolved into different breeds in each country. They were used as 'match cocks' in cock fighting for entertainment, and were common farmyard fowl across England.

Breeding and Management

The Old English Game Bantam is a bold, active and hardy breed, primarily used as pets and for exhibition. They are not as flighty or aggressive as Oxford or Carlisle Old English Game, although they will attempt to stand their ground when challenged. Their wide, well developed breast exhibits strong and prominent pectoral muscles, and carries plenty of meat for a small bird.

The eggs are very small and can vary in colour, depending on the source. Officially they should be tinted. Productivity can vary greatly, depending on the strain and the source of the stock.

Bantam Old English Game can be kept by hobbyists with limited room, and will fare equally as well on open free-range or in a shed or aviary. Perches should be of a diameter suitable for birds to grasp comfortably, to ensure feet and breastbone deformities do not occur in stock destined for a show. Colour is important in Bantam Old English Game, but above all the bird must be well balanced, broad and symmetrical at the shoulders tapering to the tail. The back should be flat, and this can be determined by running the fingers down the back; a hollow/dipped back is undesirable. The flesh should feel firm yet corky, giving the bird substance.

It is important to tame and handle birds often to ensure they remain calm yet confident in a show cage.

Varieties include the following:

- Spangle
- Black
- Black-red (partridge females)
- Black-red (wheaten females)
- Blue
- Blue duckwing
- Blue furness
- Blue-grey
- Blue-red
- Blue-tailed wheaten hen
- Brassy-backed black
- Brassy-backed blue
- Brown-red
- Crele
- Cuckoo
- Furness
- Ginger-red
- Golden duckwing
- Lemon blue
- Pile
- Silver duckwing
- Splashed
- White

Birds exhibiting muffs and/or tassels are also recognized, and can be used in any of the colour varieties.

Main Uses

- **For exhibition**
- **For crossing:** OEG bantams could prove useful for the development of new colours in other bantam breeds

Spangle Old English Game bantam cockerel with a proud and keen carriage.

Spangle Old English Game bantam pullet showing a broad breast and prominent pectoral muscles which are characteristic of both sexes.

OLD ENGLISH GAME, CARLISLE TYPE

Origin: Ancient British breed
Class: Hard feather
Colour varieties: Twenty-seven varieties (see below)
Purpose: Exhibition
Eggs: Tinted
Weight: Large fowl, cock: cocks and cockerels up to 2.94kg (6.5lb); large fowl hen: hens and pullets up to 2.5kg (5.5lb)
Ancestry: Not known
Sitter: Yes
Auto-sexing: No

History

Year created: Not known.

Originator: Various breeders wishing to exhibit Old English Game.

Breed development: The period when Game fowl were created and first domesticated is not known. The history of the Carlisle and Oxford Old English Game is closely linked and remained largely the same until the 1930s, when the Old English Game divided into two separate breeds.

The Old English Game was originally known as Game fowl prior to the banning of cock fighting in 1849, and is often referred to as the 'Pit Game'. Up until 1849 cock fighting had been popular entertainment for 2,000 years or more, for both the common man and the aristocracy. But the way in which the breed was selected for exhibition caused tension between those who wanted to maintain the Pit Game, and those who wanted desirable exhibition birds. The original Pit Game ultimately divided into two separate breeds in the 1930s, with separate standards and breed clubs including the Oxford and Carlisle clubs. Those who formed the Carlisle club were interested in maintaining the broader breasted, more horizontally backed birds for exhibiting. (D. Scrivener, 2009.)

Bantam: None exists. The OEG bantam follows the Carlisle ideal.

Young Cuckoo Carlisle type OEG. To fully appreciate the differences between Oxford and Carlisle types, mature stock should be seen at first hand.

Breeding and Management

The Carlisle type was predominantly developed for exhibiting, and still is today. They are hardy, energetic and strong. In many ways Carlisles are similar in habit and requirements to the Oxford type, but differ by being wider and more prominent across the shoulders. Their carriage is held nearer to the horizontal than Oxfords, and they are heavier and bulkier, losing much of the elegance of the true cock-fighting fowl. The body should appear heart-shaped and flat when the wings are closed and viewed from above. The body tapers well from the shoulders to the tail. The flesh is firm yet corky and they are single combed.

The Carlisle type could be used for low maintenance free-range rearing, and would benefit from plenty of running space. They can be flighty and have an aggressive temperament. Carlisle OEGs have excellent mothering instincts and are very protective of their offspring.

It is often hard to source breeding stock of both Carlisle and Oxford Old English Game, the main reason being that breeders do not wish to advertise their stock, nor do the breed clubs desire to advertise their members due to the common theft of stock. Highly prized stock and rare colours are often at risk of being stolen as they are still used illegally by some people for cock fighting as well as for breeding.

Cuckoo Carlisle Old English Game large fowl female.

Birds destined for exhibition may be dubbed – see panel under Modern Game earlier in this chapter.

Varieties include the following:

- Birchen grey
- Black
- Black-breasted grey
- Black-red (partridge female)
- Black-red (wheaten female)
- Black-splashed
- Blue
- Blue-grey
- Blue-splashed
- Brassy-backed
- Blue duckwing
- Blue furness
- Blue-red
- Blue-tailed wheaten hen
- Brassy-backed blue
- Brown-red
- Crele
- Cuckoo
- Crow wing
- Furness
- Golden duckwing
- Lemon-blue
- Pile
- Salmon-breasted blue
- Silver duckwing
- Spangle
- White

Main Uses

- **For exhibition:** Old English Game, Carlisle type is an enhanced version of the original Old English Game (Pit Game/Game Fowl)
- **For low maintenance, free-range rearing**

OLD ENGLISH GAME, OXFORD TYPE

Origin: Ancient British breed
Class: Hard feather
Colour varieties: Thirty-plus colour varieties
Purpose: Original fighting cock, now purely for show and cross-breeding
Eggs: Tinted
Weight: Large fowl cock, 1.8–2.5kg (4–5lb); large fowl hen, 0.9–1.36kg (2–3lb)
Ancestry: Not known
Sitter: Yes
Auto-sexing: No

History

Year created: Not known.

Originator: Not known.

Breed development: The period when Game fowl were created and first domesticated is not known. The history of the Carlisle and Oxford Old English Game is closely linked and remained largely the same until the 1930s when the Old English Game divided into two separate breeds.

The Old English Game was originally known as Game fowl prior to the banning of cock fighting in 1849, and is often referred to as 'the Pit Game'. Until 1849 cock fighting had been popular entertainment for 2,000 years or more for both the common man and the aristocracy. The breeders of the original Pit Game split the breed into two types in the 1930s: the Carlisle and the Oxford. Those who formed the Oxford club wanted the OEG to retain the original 'Game' type with the qualities of the Pit Game. (D. Scrivener, 2009.)

Prior to the banning of cock fighting, birds prepared for cock fights were fitted with metal spurs, known as 'gaffles' or 'goblocks'. They were made of iron, brass or silver, and were fitted tightly to the bird to prevent them coming off in a fight.

Bantam: Bantam Oxfords also exist, although rare, and they follow the large fowl type.

BELOW LEFT: Immature Brown Red Oxfords. The cockerel has not yet fully developed his hackle or wing bow colour.

BELOW RIGHT: Brown Red Oxford. Some have brown lacing on the breast to the top of the thighs.

Breeding and Management

The Old English Game, Oxford type is an agile, athletic, aggressive and flighty breed suited to low maintenance, free-range rearing. Oxfords are perhaps the fastest native breed on their feet, with Old English Pheasant Fowl, Derbyshire Redcap and Hamburg coming close behind.

The Oxford Old English Game is quite possibly the hardiest, most resilient and defensive of all native breeds. They can be used for domestic egg production, and produce medium-sized eggs of various shades from white to pale brown, although officially they should be tinted. The incorrect colour may be a sign of past crossing.

Oxfords can be used as table birds in their own right, but ideally are used for crossing with soft-feathered breeds to improve the type for table use – even though the quantities of meat produced are far from that of modern broilers. They are of the original cock-fighting type: alert and agile, close-heeled and courageous. Traditionally it is considered that a good game fowl cannot have a bad colour, and that the habit, alertness, type and balance of a bird are most important.

Birds are faced away when they come before a judge, to assess whether the specimen has the correct balance: they should have a short back with broad shoulders that taper to the tail. The back is held at approximately forty-five degrees. The breast and pectoral muscles are prominent, the flesh is firm yet corky, and they are single combed.

Old English Game would benefit from plenty of running space during the day to remain fit and agile.

Cockerels and cocks should be kept separate at all times. Young, growing cockerels can be kept together provided hens and pullets are kept out of sight. Never mix males that have never seen one another before. Some large breeds will make do with a short scrap with some minor injuries, but Oxford Old English Game are altogether more vicious and determined.

Males will defend females, and broody hens are very defensive, so beware when trying to collect eggs from beneath them. Oxfords, and particularly males, are not suitable for use near children or in confined spaces with other pets due to their sometimes unprovoked aggression.

Their housing requirements are not special, but solid divides between pens will prevent fighting through the wire mesh, and wire mesh becoming torn.

Birds destined for exhibition or the show ring may be dubbed – see panel under Modern Game earlier in this chapter. The amount of comb removed in Old English Game is down to personal preference: some prefer to leave a little, others cut back close to the top of the head.

The Oxford breed includes the muff and tassel varieties, and the Hennie Game, which is the hen-feathered Pit Game.

Main Uses

- **For exhibiting:** Oxfords can be exhibited as the true cock-fighting type

Pile Oxford cockerel.

- **For reintroduction:** Oxfords could be reintroduced into native breeds with OEG parentage, of which there are many, to restore or improve type or increase vigour
- **For low maintenance, free-range rearing**

A broody Pile Old English Game pullet sitting on nine eggs.

Dubbed comb, wattles and ear lobe of an Oxford OEG. The male shows the typical dark purple (gypsy) face colour of the Brown Red variety.

Head of a female Brown Red.

Partridge Oxford Old English Game female.

OLD ENGLISH PHEASANT FOWL

Origin: Yorkshire, Lancashire, England
Class: Light, soft feather
Colour varieties: Gold, silver
Purpose: Eggs
Eggs: White
Weight: Large fowl cock, 2.7–3.2kg (6–7lb); cockerel, 2.5–2.7kg (5.5–6lb); large fowl hen, 2.25–2.75kg (5–6lb); pullet, 2–2.25kg (4.5–5lb)
Ancestry: Gold-spangled Yorkshire Pheasant, Lancashire Mooney Fowl
Sitter: No
Auto-sexing: No

Old English Pheasant fowl cockerel with excellent colour and plenty of striping in the neck hackle.

History

Year created: Officially recognized as a unique breed in 1914.

Originator: Not known.

Breed development: The Old English Pheasant Fowl was established by keen poultry breeders to amalgamate and preserve the few remaining farmyard fowls of Northern England, including the Gold-spangled Yorkshire Pheasants and some strains of Lancashire Mooney Fowls that were not absorbed by the Hamburgh breed. (D. Scrivener, 2006.)

Bantam: There has never been a bantam variety.

Old English Pheasant fowl hen.

Breeding and Management

The Old English Pheasant Fowl (OEPF) is one of the most energetic and flighty of native breeds. They are suited to domestic egg production, and could be used for crosses to create low appetite egg-laying hybrids. The OEPF is a hardy breed that has evolved to survive in

Old English Pheasant fowl cockerel showing the laced breast feathers.

grown on separately from heavyweight breeds. As with several other primitive-type, lightweight breeds such as the Hamburgh and Scots Grey, they are particularly poor performers when hatched late in the year from August onwards, outside the natural breeding and growing season. Decreased day length decreases their desire to eat and grow, with the result that they usually die young, before or during the winter. When mature, however, the OEPF is exceptionally hardy; it is not prone to any particular ailment, and ages well.

The OEPF is supposedly larger than the Derbyshire Redcap and Hamburgh, at least as far as the breed standard is concerned. However, the breed is often below its standard weight, largely due to its rarity – it is rarer than the Derbyshire Redcap and Hamburgh – and because it is kept and bred in only small numbers by hobbyists. It is rarely seen or exhibited, and there is no dedicated breed club to encourage consistency or improvement within the breed. Therefore it is important to select for the largest and most vigorous birds when breeding.

Although traditionally a smallholder's dual-purpose breed for those that lived on harsh, low maintenance environments where evading predators and finding much of their own food was part of day-to-day living on a farmyard.

The flightiness of the OEPF may amount to short distance flights of perhaps twenty metres at a couple of metres off the ground. They will scale two-metre fencing easily and regularly, so penned birds should be netted over the top of the pen to prevent escape.

OEPF chicks are sometimes weaker than many other breeds, and would be better

Day-old Old English Pheasant fowl.

little, the breed has insignificant table value by comparison with the modern broiler. Egg-laying ability and shell quality is still good, although selection for correct egg shape is required as oval or spherical eggs are often seen in some stock. Egg colour can also vary from white to a pale brown tint. White is the correct and desirable egg colour for this traditional British breed. Any other colour would suggest past outcrossing, possibly a result of some breeders attempting to improve plumage markings or body size by using another breed.

Some OEPF will naturally develop a pale gingery ground colour, and others will develop with excessively sized black markings. These are not desirable features for exhibition or breeding. A bright, rich bay ground colour should be sought, and males should not have excessive black on their breast or on their neck hackle. The breast lacing should be clearly defined and not heavy.

The Old English Pheasant Fowl is similar in appearance to the Derbyshire Redcap, and it is important to avoid crossing the breeds together. OEPF have white ear lobes and males have a laced breast. Derbyshire Redcaps have red ear lobes and males have a solid black breast and underbody. The Redcap comb is also substantially larger than that of the OEPF.

Silver Old English Pheasant fowl have a silver-white ground colour with beetle green-black lacing to the breast of males. They probably no longer exist, but a cross with the Silver-spangled Hamburg could soon revive the silver variety. The Silver OEPF has never been numerous, and the birds that did exist may well have been absorbed by the Hamburgh breed. If times were different and less interest were shown towards preserving our native breeds, it is quite likely the Gold OEPF would have been absorbed by the Derbyshire Redcap or Gold-spangled Hamburgh. The need for a breed to be unique and with a definitive purpose is of paramount importance to ensure its survival.

Old English Pheasant fowl cock with an excellent square-fronted rosecomb with plenty of workings (spikes). Unfortunately this cock has solid black, breast and neck hackle feathers, which are not desirable.

Main Uses

- **As a dual-purpose breed**
- **For crossing:** The OEPF is suitable for crossing with productive heavyweight laying breeds to create a low appetite hybrid with a more energetic and hardy free-ranging habit
- **For low maintenance, free-range rearing:** A low appetite breed suited to roaming a farmyard or field. They are best suited to a free-ranging environment and may perch high in hedgerows or trees at night. They would be safer in a securely fenced pen or field, but if left to roam free where circumstances allow, losses to predators may be expected, although far fewer than if a large breed such as the Orpington were used

ORPINGTON AND AUSTRALORP

Origin: Orpington, Kent, England
Class: Heavy, soft feather
Colour varieties: Standardized: blue, black, buff, white. Non-standardized: see below
Purpose: Eggs, exhibition
Eggs: Pale brown
Weight: Large fowl cock, 4.5kg (10lb) min.; large fowl hen, 3.6kg (8lb) min.
Ancestry: Varies depending on colour variety
Sitter: Yes
Auto-sexing: No

Large Buff Orpington pullet owned by Gavin and Irene Keys of Essex. The pullet has a desirable consistent shade of buff all over.

Large Buff Orpington cockerel owned by Gavin and Irene Keys of Essex.

History

Year created: Black Orpingtons were the first variety and were first exhibited in 1886.

Originator: William Cook of St Mary Cray, Orpington, Kent.

Breed development

Orpington: The Orpington was developed as an attractive dual-purpose breed. Each variety was developed on different lines, although they have all been selected to one breed standard.

The original utility Black Orpington is essentially today's Australorp. The black variety was created using a Black Minorca cock crossed with black pullets derived from the Barred Plymouth Rock. The female offspring were then crossed with Black Langshan cocks. The

Orpington changed from a utility breed to one purely for exhibition shortly after its creation. Only those Black Orpingtons that made it out to Australia and preserved for their utility merit by Australians remained similar to the original type. The exhibition type Black Orpington was developed by Joseph Partington of Lytham, Lancashire, shortly after William Cook's creation.

The White Orpington, also created by William Cook, was made by mating a White Dorking cock with offspring from crosses of White Leghorn and Black Hamburgh. It is likely that White Cochins were later crossed with the resulting offspring to create the desired Orpington type.

The Buff Orpington was the third variety created by William Cook and exhibited in 1894. It was created from a mating of a Buff Cochin cock mated to pullets from a mating of Gold-spangled Hamburgh and Dorking. The Cochin first came to Great Britain from Shanghai as early as 1847.

The Jubilee Orpington was initially developed by William Cook for Queen Victoria's Diamond Jubilee and was first exhibited in 1897. They were made from dark Buff Orpingtons, Red Dorkings and Gold-spangled Hamburghs.

The original blue and cuckoo varieties were created by Arthur C. Gilbert and were both first exhibited in 1907.

Red Orpingtons were created by selecting and breeding dark-coloured Buffs by W. Holmes Hunt of Brook House Poultry Farm, Hellingly, Hailsham, East Sussex between 1905 and 1909.

The Black Orpington became purely an exhibition breed from very early on. White and buff varieties were maintained by both exhibitors and those wishing to use them for their utility merit; nowadays, however, most White and Buff Orpingtons are the profusely feathered exhibition type. Jubilee and Red Orpingtons still exist, but may now have Sussex ancestry, either because they were absorbed by the Sussex early in the twentieth century when they were less popular, and/or because new strains have been developed with the help of the Sussex. The Blue Orpington is purely an exhibition variety. The cuckoo variety still exists but may have been redeveloped later in the twentieth century. (D. Scrivener, 2009.)

Australorp: This breed was first imported into the UK from Australia in 1921, some thirty-four years after William Cook's original utility Black Orpington was exported to Australia. There was a need to protect and distinguish the utility Black Orpington (the Australorp), as maintained and improved in Australia, as distinct from the common exhibition Black Orpington. Therefore the Australorp was first standardized in the UK in 1928, and then later in Australia in 1930.

The Australorp was originally black, but can now also be found in white and blue, and there is also a bantam version. In truth, the Australorp is a variety of the Orpington breed much like a different colour, and actually a utility version of the Black Orpington with the same parentage. Despite this, the Australorp has been established as a separate breed, adding to confusion.

Bantam: Bantam Orpingtons are very popular among hobbyists, and were first created shortly after the development of the original large fowl varieties.

Breeding and Management

The Orpington and Australorp are hardy and docile breeds that are neither aggressive nor flighty. The Orpington has developed to become a purely exhibition type, with some strains retaining a productive egg-laying ability, producing medium to large pale brown eggs. The table qualities of most Orpingtons are negligible, so if a dual-purpose breed with potential as a domestic table bird is required, the Australorp would make a better choice.

Some Orpington colours such as jubilee, white and red have been absorbed by the speckled, white and red Sussex varieties. Evidence of this may come in the form of pale brown eggs which are common of those colours of Sussex.

The Orpington remains a firm favourite amongst newcomers for its calm, docile nature, soft profuse feathering and reliable brooding ability, all of which makes them an ideal breed to be kept where there are children in the family. Orpingtons are magnificent birds, particularly when seen in small flocks roaming freely; however their colours are usually poorly defined as a result of having such profuse feathering. Their soft, abundance of feather and large slow carriage make them easy to handle but more susceptible to lice and predators.

A great threat to the Orpington today is the tendency of potential owners to purchase birds from the show élite. These birds are giants, excessively feathered and slow in nature, and are reported to be very infertile. Their exaggerated size has made mating more difficult, and extensive inbreeding and line-breeding is another hindrance to their vigour. However, it would seem that low fertility and hatching rates are accepted by some breeders because although only perhaps a dozen chicks hatch from a hundred eggs, all they need is one 'golden bird' to win at a show.

Orpingtons have a deep and broad body, with short curved back, giving a concave outline. The Orpington standard states that the carriage of an Orpington should be 'Bold, upright and graceful, that of an active fowl'. However, 'graceful' and 'active' would not seem to apply to any modern Buff Orpington, as they are certainly not graceful – despite their profuse feathering, they bound along rather than glide. The Orpington's very nature is to be calm, docile and slow, so the epithet 'active' does not truly apply, particularly when compared to a renowned active, energetic and flighty breed such as the Old English Pheasant Fowl. The OEPF standard states that the breed carriage should be 'alert and active', and the two breeds are worlds apart if a comparison of habit were to be made.

When breeding Buff Orpingtons it is important to select a cockerel with a very similar shade of buff to the hens. If a dark buff bird is crossed to a pale one, a mixture of mealy coloured and patchy birds will be created.

The Australorp has retained the principles of the original Orpington created by William Cook, and is a far more productive layer and table bird than the Orpington itself. Australorps are effectively a variety of the Orpington, but have been established as a separate breed to ensure the survival of their utility qualities. Although they were created in Australia they are effectively identical to the original Orpington, therefore they are effectively an Australian strain of the Orpington breed. Currently the Australorp is very rare in Great Britain, despite having the potential to be a valuable all-rounder.

Orpingtons and Australorps require secure fencing and housing to protect them from predators; unlike the Old English Pheasant Fowl or Old English Game, they stand no chance of running away from a predator. They are a suitable breed for free range and will forage, but they are often just as at home in a small pen.

To keep birds and their delicate plumage in prime condition for exhibition, a small sheltered aviary would be ideal. When exhibiting Buff Orpingtons it may be necessary to keep them out of the sun where possible in order to maintain a consistent buff shade.

Orpington cockerels will get along well with one another right into maturity, provided they are kept together when young. This is the case with many breeds, but Orpingtons are particularly docile. The introduction of a new Orpington cockerel into an established breeding pen containing another male may result in a couple of short scraps before they accept each other's company. Of course this is not always the case, but generally they do not have the energy for long drawn-out battles.

Large Black Orpington female, owned by Andrew Macredie of Nottinghamshire.

Bantam Chocolate Orpingtons. (Courtesy Andrew Macredie)

Large Blue Orpington female. (Courtesy Andrew Macredie)

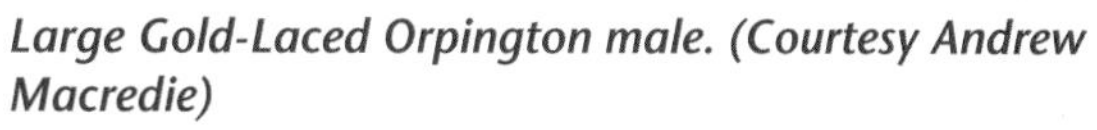

Large Gold-Laced Orpington male. (Courtesy Andrew Macredie)

Large Splash Orpington female. (Courtesy Andrew Macredie)

Large Gold-Laced Orpington female. (Courtesy Andrew Macredie)

Large Buff Orpington cockerel: a dark buff with tighter feathering than usual.

Large Buff Orpington pullet, poorly coloured and inconsistent. Mealy buffs like this come from crossing dark buff birds with pale buff ones. To ensure the offspring have a consistent shade, ensure that both parents are a similar shade of buff.

Some of the Orpington colours not yet standardized by the Poultry Club of Great Britain but which still exist include the following:

- Blue-laced
- Chocolate
- Cuckoo
- Gold-laced
- Jubilee
- Lavender
- Lemon cuckoo
- Partridge
- Silver-laced
- Splash
- Red

Main Uses

- **For exhibition:** This is the main purpose of Orpingtons in Great Britain
- **As a sitter (broody):** Orpingtons and Australorps are probably one of the most reliable native sitters
- **As a table breed:** The Australorp is suitable for table use. However, Orpingtons do not fill out quickly enough, or even at all, to be worthwhile table birds
- **As a productive egg layer (both the Orpington and the Australorp):** Only some strains of Orpington are reliable layers of large eggs, and it is best to avoid strains that are persistently used in the exhibition scene. Australorps are the original utility Orpington and are the breed of choice for utility purposes

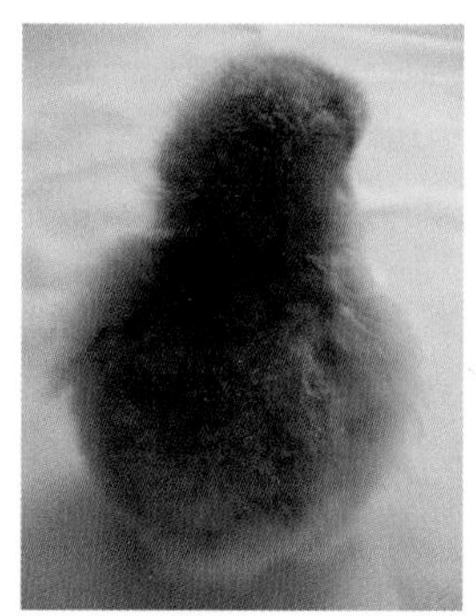

Day-old large Buff Orpington.

ROSECOMB BANTAM

Origin: Great Britain
Class: True bantam
Colour varieties: Black, blue, white, black-breasted red, birchen, columbian
Purpose: Exhibition
Eggs: White or cream
Weight: Cock, 570–620g (20–22oz) weight; hen, 450–510g (16–18oz)
Ancestry: Not known
Sitter: No
Auto-sexing: No

History

Year created: Not known.

Originator: Not known.

Breed development: It is believed that bantam fowl had been imported into Europe from Bantam, Indonesia, since the fifteenth century. These variable miniature fowl became established as a unique breed in various European countries, including Great Britain. The history of how they were selected or crossed is not known, although crosses with Hamburghs are likely to have been made to improve the Rosecomb as an exhibition breed.

The first record of the existence of Black Rosecombs was of a flock owned by John Buckton in 1483, who owned the Angel Inn at Grantham, Lincolnshire. The Rosecomb could never be a useful utility breed due to its small size, so it has been gradually developed since the first poultry shows of the mid-nineteenth century to become a specialist exhibition breed. (D. Scrivener, 2009.)

Breeding and Management

Rosecombs are suited to domestic keepers who only have room for a small aviary and require a small, attractive breed. Egg-laying ability is completely dependent on the strain and the source of stock. The breed is not a utility one, so egg-laying ability cannot be expected to be high or reliable. Rosecombs are not particularly flighty or aggressive,

Black Rosecomb male. (Courtesy Jane Eardley)

Black Rosecomb female. (Courtesy Jane Eardley)

but should be handled regularly to ensure they remain calm and display themselves well when in a show cage. They are usually exhibited within their first year. Because they have such a short showing career, they are often hatched as early as February to be exhibited in November or December.

Rosecombs are exclusively for the exhibition scene, and should be maintained and bred to the highest standards to ensure the breed retains the near-perfect attributes that have been achieved so far. To ensure that plumage remains unbroken and lobes remain undamaged, specimens destined for a show should be kept isolated in a small, sheltered pen. Solid sides rather than wire or weldmesh will help to prevent wear and tear to lobes and plumage. Ideally the pen should be situated where the birds cannot be startled by predators or cats.

Perches should be situated close to the floor and far enough away from the wall to ensure the long tail has plenty of room; this will help to prevent the sickles becoming worn or broken. Perches must be small enough in diameter to ensure that a bird's feet can grasp it comfortably, to avoid adverse effects: see panel.

Only keep a cockerel with pullets or hens for breeding purposes because a treading cockerel may well damage the ear lobes, comb or plumage. To ensure consistent high quality, double mating is often utilized.

To understand what to look for, speak to an exhibitor of the Rosecomb; and for those intending to take exhibiting seriously, it is advisable to buy direct from an exhibitor rather than at an auction, to ensure you start with worthwhile stock. Furthermore take up any opportunity to see at first hand how different breeders select and maintain their stock.

Rosecombs must be correctly proportioned and correctly coloured. The back should form a smooth, sweeping curve from the head to the tail. The tail must not be held at a sharp angle to the back, but must curve gracefully from the back to the

PERCHES

A perch must be small enough in diameter to ensure that a bird's feet can grasp it comfortably, to avoid adverse effects: if it is too wide, birds may become duck-footed, where the hind toe becomes incorrectly positioned. Also, after some time a bird that has used a perch which is too large or uncomfortable to grasp may develop a crooked breastbone; this can sometimes be felt when the bird is held, and is a fault in any breed.

Day-old Black Rosecomb. (Courtesy Jane Eardley)

tip. Black birds have black legs and feet, while the white variety has white legs and feet. The ear lobes are white. The comb is small, with even workings and a single straight leader that runs level with the top of the comb; the quality of the comb and earlobes are important attributes of the breed. The tail is very large, held well behind the body; it must not be held too high.

The ear lobes in the male must not be smaller than 1.9cm (¾ in) or larger than 2.2cm (⅞ in). In the female, lobes should not be larger than 1.6cm (⅝ in). Despite these standard requirements, there is a great and accepted tendency for ear lobes to be larger in both males and females.

Main Use

- **For exhibition:** This breed is developed purely for exhibition. The rose comb and white ear lobes are the most important features, potentially taking the greatest number of points at shows

A creased wattle and an undesirable dished ear lobe with slight scabbing and reddening. (Courtesy Jane Eardley)

A male Rosecomb with a poor leader. The leader should be tapered and level with the top of the comb. The comb has a square front and has plenty of rounded workings on the surface. (Courtesy Jane Eardley)

SCOTS DUMPY

Origin: Scotland
Class: Light, soft feather
Colour varieties: Black, cuckoo and white; standardized colours of other breeds are acceptable
Purpose: Eggs
Eggs: White
Weight: Large fowl cock, 3.2kg (7lb); large fowl hen, 2.7kg (6lb)
Ancestry: Not known
Sitter: Yes
Auto-sexing: No. Possible in the cuckoo variety, although reliability will vary depending on strain and selection

History

Year created: Not known, an ancient breed.

Originator: Not known.

Breed development: The history of the Scots Dumpy, like the Scots Grey, is largely unknown. They were originally a crofter's breed used for eggs, meat and as broodies, but have now become primarily an exhibitor's breed. Scots Dumpies have been known by several nicknames in the past, including Bakies, Creepies, Daidlies, Go-laighs, Hoodlies, Kities and Stumpies, their homeland being a band across Scotland from the Isle of Arran in the west to Aberdeen in the east. They were first exhibited in 1852 by Mr Fairlie of Cheveley Park, Newmarket, Suffolk, England. The Scots Dumpy fell out of favour in the fancy during the middle of the twentieth century, and was maintained by only a few dedicated breeders in Scotland.

Large Cuckoo Scots Dumpy cockerel with a single dose of the barring gene, resulting in wider black bars and an overall darker shade.

Large Cuckoo Scots Dumpy pullet. Cuckoo females remain a consistent colour: they can only carry a single dose of the barring gene.

A revival of interest in the breed, and the increased inbreeding of the few that remained in the UK, encouraged breeders to import eggs in 1973 from a strain kept by a Mrs Violet Carnegie, who had emigrated to Kenya. Since then the breed has gone from strength to strength, and the Scots Dumpy club was re-formed in 1993. Currently, black and cuckoo are the main colours used by exhibitors; white is rarely seen. A blue variety also exists, but is in the hands of only a few exhibitors.

Prior to the first exhibit of Scots Dumpies in 1852 their original colours included silver-hackled black, gold-hackled black and speckled reds. (D. Scrivener, 2006.)

Bantam: The bantam was probably created using undersized large fowl Scots Dumpies and bantam Scots Greys. Mr J. Craig of Dreghom, Ayrshire, Mr J. Lindsay of Arrochar and James Garrow of Loanhead, Midlothian, were responsible for their creation. (D. Scrivener, 2006.)

Breeding and Management

The Scots Dumpy is an active and friendly breed that is rarely aggressive or flighty. They are very hardy, and are productive layers of medium-sized white to pale brown eggs, depending on their strain; officially these should be white, but pure white eggs are rarely seen.

Scots Dumpies are easily handled, have excellent mothering instincts, and are suited to low maintenance, free-range rearing. Most significantly, some specimens carry what is known as the 'creeper' trait (*see* panel below), meaning that they have very short legs. This is a feature which has been encouraged in the breed as far back as it has been recognized, and may have been a useful attribute when the Scots Dumpy was originally kept on crofts since short legs would have allowed it to roam through heather and scrub without being seen by airborne or ground predators. However, short legs are not ideal for cold, wet or boggy ground, which may adversely affect the bird's overall health and condition.

Despite their short legs, Scots Dumpies are not deterred from perching and will still manage to perch at least two feet (60cm) from the ground.

The long-legged birds, and in particular the cockerels, can and should be used for breeding, since the longer legs help with mating and will eliminate the adverse effects of the creeper gene.

The Dumpy is an active breed, but neither short- nor long-legged birds are energetic and fast in the manner of the Old English Pheasant Fowl or Old English Game, so they are more susceptible to predators, which means that secure fencing and housing is essential.

Dumpy chicks feather up quickly and do so slightly faster than most other native breeds. Any white plumage seen in young black chicks may be a sign of past crossing between

THE CREEPER TRAIT

The short legs of some Scots Dumpy specimens are caused by what is known as the creeper trait; however, this trait can have a seriously adverse effect when two short-legged birds are mated together, because 25 per cent of the offspring will receive a double dose of the gene, which results in chicks dying in the shell.

Statistically, a short leg × short leg mating can produce the following ratio of chicks:

50 per cent short-legged
25 per cent long-legged
25 per cent dead in shell

A long leg × short leg mating will eliminate the adverse nature of the gene and produce the following results:

50 per cent short-legged
50 per cent long-legged

the colours, but does not necessarily persist into adulthood. According to the standard there is no fixed plumage colour, which for exhibitors allows the creation and showing of a vast array of varieties, if they have the enthusiasm to create them.

Cuckoo Scots Dumpy males can vary in shade due to the barring gene. A single dose of the gene produces a bird genetically impure for the barring gene with wide black bars, resulting in a darker shade. A double dose of the gene produces a bird genetically pure for the barring gene with narrow black bars and an overall pale cuckoo colour. Females only carry a single copy of the gene, so are always of a darker shade. The variation in some male stock is most likely the result of past crossing with Black Scots Dumpy females. Such a crossing causes the male offspring to carry a single dose of the barring gene. These single barred males later go on to breed with Cuckoo females to produce pure and impure Barred males, Cuckoo females and Black females. A genetically pure male, one that carries two copies of the gene, is required with Cuckoo females to breed 100 per cent

Large long-legged Black Scots Dumpy cockerel.

Large long-legged White Scots Dumpy cock: some black ticking can be seen, and a small cuckoo feather is present on the bird, indicating that this particular white bird may have been created by crossing white with black or cuckoo stock. This means that he will quite possibly throw up black and cuckoo offspring at a later date.

Large Black Scots Dumpy pullet.

Large Cuckoo Scots Dumpy cockerel with a double dose of the barring gene, resulting in narrower black bars and an overall paler shade.

Large Cuckoo Scots Dumpy cock with many of the common colour faults seen in Cuckoos, including gold, solid black and sometimes solid or predominantly white feathers.

Cuckoo Scots Dumpies. However, birds with nearly even bar widths, as seen in the darker form, are preferred to meet the breed standard when exhibiting, therefore careful selection is required over many generations to achieve this uniformity and to reduce the occurrence of solid black feathers.

For exhibiting, the shank should be no longer than 3.75cm (1.5in). Long-legged Scots Dumpy males are useful for breeding.

To create new bloodlines within the Cuckoo variety, it is possible to cross a Cuckoo male that is genetically pure for the barring gene with Black females. The resulting pullets will be Cuckoo and can be used for breeding with Cuckoos. The male offspring will also be Cuckoo but will only carry a single copy of the barring gene, therefore being impure for the barring gene, and should not be used for breeding.

Bantam Scots Dumpies are very rare, but there may be some confusion among hobbyists as to whether their birds are large

Head of a Cuckoo Scots Dumpy cockerel.

fowl or bantam. Many are even unaware of the existence of a bantam version, the reason for the confusion being that the bantam is often larger than is required by the breed standard.

The Scots Dumpy as a whole is a scarce breed, so it is not hard to appreciate that some confusion exists.

Main Uses

- **As a dual-purpose breed:** For a lightweight breed, Scots Dumpies produce a reasonable amount of meat by traditional standards, as much as, for example, the Norfolk Grey. They are also reliable layers of medium-sized eggs
- **As a sitter (broody)**
- **For low maintenance, free-range rearing:** Scots Dumpies could forage for part of their diet in scrub, hedgerow or wooded areas, their short legs and semi-tight plumage helping them to traverse such environments easily

Bantam Black Scots Dumpy cockerel showing the desirable glossy green sheen that should be found on both sexes.

SCOTS GREY

Origin: Scotland
Class: Light, soft feather
Colour varieties: Barred
Purpose: Eggs, though traditionally this was a dual-purpose breed
Eggs: White
Weight: Large fowl cock, 3.2kg (7lb); large fowl hen, 2.25kg (5lb)
Ancestry: Not known
Sitter: No
Auto-sexing: Yes

History

Year created: Not known.

Originator: Not known.

Breed development: The history of the Scots Grey is largely unknown. The Scotch Grey Club was formed in December 1885 in Edinburgh, though it was later renamed the Scots Grey Club in about 1923. Nothing is known about the development of the Scots Grey prior to the formation of the club – probably it was a long-established breed used by crofters in Scotland for eggs and meat. (D. Scrivener, 2006).

Large Scots Grey cockerel.

Large Scots Grey hen, five years old. Entirely black feathers such as on the breast of this specimen are undesirable but do commonly occur.

The Scots Grey may be derived from the same stock as the Scots Dumpy, or it may even be related in some way to the Game fowl, known as Old English Game since the banning of cock fighting in 1849. These breeds are among the oldest natives, also the Scots Dumpy originates from the same area, and they all have similar characteristics to the Scots Grey.

Bantam: The history of the creation of the bantam is also largely unknown; the first known exhibitor was Mr Mitchell of Paisley who had kept them as far back as 1866. (D. Scrivener, 2006.)

Breeding and Management

The Scots Grey can be highly recommended for its hardy nature, its suitability for free-range

Bantam Scots Grey cock.

Bantam Scots Grey hen.

rearing and for domestic egg production. However, despite the Scots Grey being traditionally used as a dual-purpose breed, the breast meat is limited.

The Scots Grey is energetic and active but not as quick as the Old English Pheasant Fowl, the Derbyshire Redcap or the Old English Game. It is sometimes aggressive but not particularly flighty, although it is still capable of scaling a two-metre fence. It does not require any special housing, and will survive under low maintenance, free-range management. Scots Greys are ideally suited to large pens where plenty of running room and grazing is provided, although they can be reared in more confined spaces; however, it may be necessary to put netting over smaller pens in order to contain them.

They are a hardy breed but must be bred and hatched early in the breeding season to ensure chicks develop well before winter. Birds hatched from August onwards often do not survive the autumn and winter. Outside the natural breeding season, reduced day length and reduced temperature limit the growth rate of many lightweight breeds, and chicks and young birds often become lethargic and die. If late-bred birds do survive the winter they rarely attain the correct size and quality in time for the breeding season. The same can be said of other light breeds, such as Old English Pheasant Fowl, Hamburgh, Marsh Daisy and Derbyshire Redcap. Otherwise birds age well and are not particularly susceptible to any complaint.

Egg shell quality varies slightly, so it is important to select breeding stock carefully in order to maintain shell quality. Eggs are of medium size and should be white, but are often slightly tinted. The laying ability of the Scots Grey is as good as other native lightweight breeds, including the Old English Pheasant Fowl and the Derbyshire Redcap.

There is a great deal of consistency among Scots Greys, perhaps due to their great antiquity and because of continued careful selection over many centuries, the Scots Grey, as far as is known, never yields black or self-coloured offspring – which may sometimes occur with other barred breeds.

Bantam Scots Greys are often heavier than the weight limits stated by the standard, so

Large Scots Grey cock with characteristic long tail.

The wing of a large Scots Grey cockerel showing how the shade should be consistent throughout with bars running straight across each feather. Barring should not be as precise as that of the Barred Plymouth Rock, and must not be imprecise or blurred like a Cuckoo.

it is necessary to select stock carefully from each generation to retain the desired weight. It is important not just to select the biggest or strongest chicks, because the aim is to maintain the bantam as a miniature fowl. Bantam chicks are surprisingly vigorous, and both large fowl and bantam chicks feather up quickly. In the bantam version there is a tendency for the wings to point down slightly as they strut about, whereas ideally the wings should be held up close to the body.

The barred colour of the Scots Grey is not often fashionable, with the result that the breed may become less numerous and more inbred. Where possible, the hobbyist should breed them in large numbers early in the year and maintain a large flock, and select or let nature select (depending on the type of management) the strongest birds of the large fowl. Do not try to look for new blood on a yearly or bi-yearly basis as you will probably struggle unless you live in Scotland

Black feathers seen on this female are a fault, and should be avoided.

where most breeders currently reside. Inbreeding will be necessary, but the adverse effects will be negligible if birds are kept in large enough numbers. Trying to source new blood on a regular basis will soon exhaust the supply.

Legs should be white, sometimes with black markings. Bantam Scots Greys often have black markings, but they are less commonly found on the legs of large fowl.

Day-old large Scots Greys. This is an auto-sexing breed: the male chick can be seen on the left, the female on the right. The reliability of auto-sexing chicks may vary depending on the strain, but could be improved with selection.

Main Uses

- **As a dual-purpose breed:** The Scots Grey is a useful breed for domestic egg production, however the breast meat is limited, certainly by modern broiler standards
- **For crossing:** The Scots Grey could be used instead of the barred Plymouth Rock in the creation of an auto-sexing breed
- **For low maintenance, free-range rearing:** It is an energetic forager with a low appetite

SEBRIGHT

Origin: Hertfordshire, England
Class: True bantam
Colour varieties: Gold, silver
Purpose: Exhibition
Eggs: White or cream
Weight: Cock, 620g (22oz); hen, 510g (18oz)
Ancestry: Uncertain; possibly Nankin, Hen-feathered Game, Gold-spangled Hamburgh, White Rosecomb
Sitter: No
Auto-sexing: No

History

Year created: The original Sebright Bantam Club was formed in 1812.

Originator: Sir John Sebright, 7th Baronet of Besford of Beechwood Park estate, Markyate, St Albans, Hertfordshire.

Breed development: The breed was developed purely along decorative lines sometime around the beginning of the nineteenth century. Sir John Sebright set up a Sebright Bantam Club as a private affair to cater for those interested in the breed. Membership was originally by invitation only, but this later changed and the current club is run like any other. (D. Scrivener, 2009.)

Breeding and Management

The Sebright is a very attractively marked breed, bred purely for exhibition since its creation. It is classified as a true bantam, meaning there has never been a large fowl counterpart. It is calm and easy to handle, and is neither flighty nor aggressive despite its small size and light build. The breed is officially spelt 'Sebright': any other spelling, such as 'Seabright', is incorrect.

The Sebright has always been highly inbred even since the early years of the breed's development: to ensure that the markings and type were perfected would have required much inbreeding. The reliance on using birds less than a year old for showing has meant there is a constant need to breed from the best of the previous year's stock to produce new stock for the current year. This has increased the rate of inbreeding, which some regard as the cause of poor fertility and hatching rates.

Older birds are not usually suitable for showing as they may develop slight colour imperfections after moulting, and are also more likely to have accrued some injury or unsightly mark such as scabbing or scarring to the comb or wattles in males. In many other breeds, older birds can be more suitable for showing, but this often depends on how the particular variety matures.

The Sebright requires a high level of perfection in order to succeed at shows, including excellent general condition. It is

Head of a male Sebright: the rosecomb is square-fronted, with the top surface covered in fine workings without a hollow surface. Ideally the leader should turn upwards slightly.

ABOVE: Silver Sebright female showing the attractive bold habit with a prominent breast and low wings. Owned by Robin Creighton.

BELOW: Silver Sebright female with a fully laced breast. Owned by Robin Creighton.

therefore essential that plumage is clean and undamaged, and that there is no sign of lice or mites, or of scabbing, scarring or injury to the head. For birds to look their best they must be calm and confident in the show cage, so it is important to handle and hand feed them so they are tame for showing purposes; a confident carriage and walk are also essential.

The breed type must be correct. The Sebright has a confident strutting carriage with a prominent breast and wings that are carried low. The males are hen feathered, meaning they do not have sickles or pointed neck and saddle hackles. The ground colour of feathers must be pure, and each feather should be slightly rounded at the tip with a single, sharply defined lace extending the entire perimeter of the feather. The tail should be full and carried high, and the main tail feathers should be well spread. Silver and gold are the only standardized colours. Citron is another colour occasionally seen in Great Britain but it is not standardized, and is a pale version of the gold variety.

It is also necessary to be aware of common physical faults that can affect any breed, which may not hinder a utility bird but are undesirable in a show bird. For example, perches that are too wide can cause adverse physical effects (*see* panel on Perches earlier in this chapter).

To maintain birds in prime condition cage them individually, or keep them in small numbers in small runs, preferably with solid sides to prevent wear and tear or damage that could be caused against wire or weldmesh sides. A roof will keep litter or soil dry and prevent plumage becoming dirty.

Only introduce a cockerel when breeding is required. Being an ornamental bantam breed the Sebright fares better if it is kept with only bantam fowl, rather than having to compete with large fowl.

The Sebright is also genetically susceptible to Marek's disease. Those that survive the disease

Lacing should follow the entire perimeter of each feather. Getting this right in the tail requires careful selection.

become carriers and should not be used for breeding.

Main Uses

- **For exhibition:** This breed has been developed purely for exhibition. The quality and extent of the lacing on each feather is of primary importance, but type and ground colour are also very important. Those wishing to breed the Sebright should seriously consider studying the breed standard closely, and should hatch as many as possible each year to select the best and to maintain a high standard of perfection within the breed
- **As a family pet:** The Sebright could be kept as a pet in a small back garden or aviary. It is a highly attractive breed that lays small eggs. Their relatively calm and easy to handle nature along with their small size and attractive colour would make them a popular choice for those who have children. They can often be found for sale at auction in both standardized colours
- **Suitable for high density residential areas:** For those with limited room or who require only a small number of eggs

Wing of a Silver Sebright with clear lacing around the perimeter of each feather, and no smuttiness on the silver-white of the feather.

SUFFOLK CHEQUER

Origin: Suffolk, England
Class: Bantam
Colour varieties: Barred
Purpose: Exhibition
Eggs: Light brown or cream
Weight: Cock,1.25–1.5kg (2lb 13oz–3lb 5oz); hen:1–1.25kg (2lb 3oz–2lb 13oz)
Ancestry: Barred Plymouth Rock bantams, brown hybrid hen
Sitter: No
Auto-sexing: They do not exhibit any obvious or reliable auto-sexing ability despite being a barred breed

Suffolk Chequer male, showing the correct stance whereby the neck and tail create a 'U'-shaped negative space when alert.

History

Year created: The Suffolk Chequer was developed between 1995 and 2010. Application for standardization was made in March 2011.

Originator: Trevor Martin.

Breed development: Trevor Martin lived in Finningham, near Stowmarket in Suffolk for twenty-three years, and it is here that he began the creation of the Suffolk Chequer. In 1994 Trevor was given a Barred Plymouth Rock bantam cock; its bloodline had originally come from Professor Geoff Parker. Another friend gave Trevor a small brown hybrid hen with hints of dark pencilling and a black tail to keep the Plymouth Rock bantam company, and in 1995 Trevor hatched one male and one female from this pair.

Trevor felt that the shortness of tail in the Plymouth Rock did not give the breed a finished look, and decided on a project to create a bantam-only breed that resembled the Barred Plymouth Rock in all respects but with a much larger tail. The first hybrids exhibited a larger tail, but to improve the colour and maintain the Plymouth Rock type, Trevor decided to cross the cockerel from the first mating back to a barred Plymouth Rock bantam hen. The bantam hen was sourced from Tom Newbould's exhibition bloodline, and the two were crossed.

From that point on Trevor has selected barred birds with the best developed tails, paying less attention to the precision of the barring,

Suffolk Chequer female owned by Trevor Martin.

although he has tried to select birds where possible with the cleanest margin between the black and white bars. By 2010 the Suffolk Chequer was breeding 100 per cent true and a proposed standard was written, and in 2011 the breed was put forward for standardization.

The Suffolk Chequer exists as a small fowl only and is a unique breed. Despite this, it may cause the Poultry Club of Great Britain (PCGB) some problems in determining its classification due to the PCGB's over-simplified rules. Trevor believes they should not be classed as a true bantam as they are a soft-feathered, heavy breed, being substantially heavier than any existing true bantam breed. One possible way of ensuring the breed is classified as a soft-feathered heavy would be to have a large fowl counterpart; however, Trevor would prefer that the breed existed only as a small fowl.

Suffolk Chequer female, front view.

Breeding and Management

The Suffolk Chequer is an active bantam with a heavy and muscular body. Birds must maintain a compact and strong body, yet remain within the proposed weight limits. Males should exhibit a 'U'-shaped negative space between the neck and tail: this is most obviously seen when they are alert and strutting around, often in a proud and defiant manner.

The preciseness of barring in males and females is not of primary importance and should not be perfectly defined as in the Barred Plymouth Rock; thus breeders should not ignore the desired type of the breed or the fullness of the tail in favour of colour markings, particularly as it may prove difficult to perfect all of them – for example the barring on a long and fast-growing tail will be broader than that on the main body. Therefore trying to maintain consistent bar width across the whole body would prove very difficult.

The tail is one of the main focal points of the breed and should be as full as possible. There should be as many as five or six main tail feathers on each side, and each one should be rounded and blunt at the tip. These feathers should be accompanied by plenty of side hangers. The two sickles on the male have a slight curve and should not greatly outreach the main tail feathers. Ideally the tip of the sickles should be in line with the rounded ends of the main tail feathers. Twisted tail feathers have been seen, although rarely, in some stock, and to ensure these do not reoccur in future it would be prudent to avoid breeding birds that have them.

At a glance the Suffolk Chequer may appear to resemble the bantam Scots Grey. Fortunately they are different in many ways. In the first instance Suffolk Chequers are heavier and have more substance when handled than bantam Scots Greys. Suffolks are also more easily handled, not being as energetic or lively as the Scots Grey. Suffolks have yellow legs, sometimes with faint red coloration on males and greyish brown to the front of the shank on females, whereas Scots Greys have white legs,

Suffolk Chequer female, rear view. The height of the main tail should be about level with the top of the head.

often marked with black. The wings on Suffolks can be carried fairly low, whereas Scots Greys should have their wings tucked up closely to the body. Suffolk Chequers produce slightly larger eggs than the Scots Grey and a meatier carcass. These good utility properties could make the Suffolk Chequer a useful breed for crossing to create a productive yet low appetite hybrid for domestic use.

Due to the ongoing inbreeding that has been required to create the breed (although Trevor has only crossed brother to sister once), hatchability is lower than would normally be expected; often the chicks will hatch a day later

Wing of a Suffolk Chequer hen showing the desirable degree of precision in the barring: this is not so precise as the Barred Plymouth Rock.

A selection of Suffolk Chequer day-old chicks; there is no obvious auto-sexing ability.

than usual, and those that do hatch late may on occasion have crooked toes. This happens because they have taken in the remaining yolk and have started to breathe for themselves whilst inside the egg, and with little room inside the egg for further growth the toes become distorted against the inside of the shell. Only by breeding in large numbers and spreading the breed far and wide across the country will genetic variability begin to occur and hopefully such problems can be avoided.

Suffolk Chequers are suitable for hobbyists of all ages and experiences, but those taking up this new breed should appreciate that each generation must be carefully selected in order to maintain the desirable carriage and tail that are characteristic of the breed. Lack of devotion to continued selection and maintenance will prevent the breed maintaining its role as an exhibition breed. With the general lack of interest in barred and indeed black-coloured breeds, it is only as a show bird of high standards that the Suffolk Chequer will survive. To survive it must compete with many other barred or cuckoo bantams, so it is of paramount importance that it remains unique and distinctive.

The Suffolk Chequer does not require any special housing, but birds would benefit from being kept in individual cages or in small numbers if they are destined for a show. Carriage and type are important features of the breed, being worth as many as forty points at a show. A small diameter, centrally placed perch allowing claws to grapple comfortably will prevent deformity to the breastbone; it will also ensure the tail has plenty of room to remain in prime condition. A pen or aviary with plenty of running room and head room will ensure that birds keep fit, and that they maintain and exhibit their characteristic 'U' shape.

It is important to handle birds regularly and tame them with food: by doing so they will be more comfortable and relaxed at a show, and will display themselves more naturally.

Main Uses

- **For exhibition:** The Suffolk Chequer has been specifically bred for exhibiting with the emphasis on type in males, and the length of the sickles and the quality and abundance of the tail feathers
- **For people in high density residential areas:** This breed is suitable for those with limited room requiring birds with utility potential

SUSSEX

Origin: East Sussex, England
Class: Heavy, soft feather
Colour varieties: Brown, buff, coronation, light, red, silver, speckled, white
Purpose: Meat, eggs
Eggs: Tinted
Weight: Cock, 4.1kg (9lb) min.; hen, 3.2kg (7lb) min.
Ancestry: Uncertain, perhaps Dorking, Old English Game, local fowl of Sussex and Light Brahma
Sitter: Yes
Auto-sexing: No

History

Year created: The Sussex Poultry Club and breed standard was formed in 1903.

Originator: Not known.

Breed development: The Sussex is a type of four-toed fowl of originally various non-standardized colours, used by farmers of mainly Sussex as a table bird for the London market. The formation of a recognized and standardized breed was encouraged by Edward Brown, a leading poultry expert at the end of the nineteenth century. He was concerned that the type of fowl developed in the region would die out if it were not properly recognized. The breed itself was a 'work in progress' prior to the formation of the breed club. When the breed standard was accepted the first three colour varieties were light, 'red and brown' and speckled.

Buff Sussex were created by several breeders, including a Mr John Raine who developed his strain between 1918 and 1920 using a Buff Orpington cockerel and Light Sussex hens.

The Cuckoo Sussex was created by Richard Terrot by crossing Cuckoo Malines with Light Sussex, followed by back crossing again to Light Sussex. Unfortunately the Cuckoo Sussex no longer exists.

The large fowl Coronation Sussex was developed in time for the coronation of King Edward VIII in 1936. Currently both large and bantam Coronation Sussex can be found, although they are particularly rare.

The Silver Sussex was created by Captain Ellis Duckworth of Merriewood, New Dome Wood, Copthorne, Sussex. The variety was recognized

Large Light Sussex cock.

Large Light Sussex hen.

by the Poultry Club of Great Britain in 1948. (D. Scrivener, 2009.)

Bantam: The bantam Light Sussex was developed from 1916 by Mr Fred Smalley at 4 Blackheath Park, London SE3. The bantam Sussex can now be found in all existing large fowl colours and, depending on the variety, these are often more common than their large fowl counterpart. (D. Scrivener, 2009.)

Breeding and Management

The Sussex is an exceptional dual-purpose breed and was traditionally the most productive layer of all British breeds. Hardy and an active forager, it is neither flighty nor does it have an aggressive temperament. Today it is primarily used for domestic egg production, producing large eggs from near white to pale brown in colour. Officially the eggs should be tinted, but due to the past amalgamation of various Orpington colours into the breed the red, speckled and white varieties of Sussex often lay pale brown eggs. Egg-laying ability varies, depending on strain and colour variety.

The Sussex has a graceful carriage with a long, broad, flat back; the tail should be held at 45 degrees in the male and 35 degrees in the female. The legs and feet are white, often with some pink-red shading along the shank. The face, comb, earlobes and wattles are red in all varieties. The chicks are usually quick to feather and are vigorous, mainly due to their continued popularity. The Sussex is not suited to low maintenance rearing, but eagerly takes up the opportunity to forage for part of its diet.

The breed has become divided into exhibition and utility types. Birds derived from exhibition strains are larger bodied with more profuse feathering around the thigh, but are usually said to be far less productive and less vigorous. Utility strains lack the unnecessary fluffiness and often have less accurately marked plumage, but they retain more useful utility traits. The Brown Sussex was developed purely as an exhibition variety, so do not expect high productivity.

The Light Sussex is the most common of the Sussex breed and is believed to have been created by crossing the Light Brahma with the existing Sussex fowl. The typical black neck striping is now seen on all Sussex varieties, which suggests that either Light Sussex or Light Brahma have been bred with the early versions of the other colours to ensure consistency throughout the breed.

The Light Sussex played a major role in the commercial and domestic poultry industries right up to the 1960s, and along with the Leghorn, Rhode Island Red and North Holland Blue was perhaps one of the most important table birds and egg-laying breeds in Great Britain. It became increasingly important for use in sex-linked crosses with Indian Game, Leghorn, Rhode Island Red, Wyandotte and Faverolle for egg or meat production. Although the breed makes an exceptional table bird, the breast meat does not develop quite so quickly as the Indian Game or Ixworth. Also, by comparison with the modern broiler they produce less meat – but the benefit of keeping a Sussex rather than a modern broiler is that it is a breed of substance, it is hardier, and not prone to any particular complaint, and is an iconic type and colour.

Juvenile Light Sussex often have black feathering exposed through the white over the whole body, which makes the bird look 'unfinished'. By maturity, however, this moults out, leaving a bird with a black neck hackle edged with white, a black tail, and black on the blade of the primary wing feathers. The feathering over the remainder of the body should be pure white throughout, and there should not be any black markings or smuttiness on the back. Nor should the black neck hackle extend far on to the back, as this makes the back look shorter and is generally considered to be undesirable.

Furthermore a bird that exhibits a white margin around the entire edge of each hackle feather is often hard to find, and many hackle feathers have the black striping persisting right to the tip. This is a common feature, and selecting for an entire white margin is only perfected for the show bench; it should not be dwelled upon if utility merit is to be maintained.

EYE COLOUR IN ALL VARIETIES

Eye colour in the Sussex varies according to the variety:

- Brown Sussex = brown or red
- Buff Sussex = red
- Coronation Sussex = orange
- Light Sussex = orange
- Red Sussex = red
- Silver Sussex = orange
- Speckled Sussex = red
- White Sussex = orange

In the speckled variety each feather should be tipped with an even size white spot, and a narrow, glossy black bar should divide the white from the red/mahogany. However, it is unlikely that a bird with a white spot at the tip of each and every feather will ever be found. The colours of each feather should not mix, and ideally there should not be any dark peppering on the ground colour. The under-colour is slate and red.

The White Sussex is an inconspicuous, plain white bird, although at one time it had the potential to be a highly useful commercial breed. Like the Light Sussex, when plucked the carcass is left without any undesirable dark stubs on the breast. Unfortunately, due to the fashion for colourful show birds and because it does not have the typical Sussex markings, it has fallen by the wayside to the point where it is now particularly rare. The Light Sussex remains by far the most common variety.

SEXING CHICKS BY DOWN COLOUR

Sex-linked crosses using the Sussex were once commonly used to ensure that day-old chicks could be sexed by down colour. By separating the males early on a farmer could either reduce his feed bill by disposing of them straightaway, or they could be grown on for the table. Genetically silver plumage is dominant, so crossing a silver male with a gold female results in all silver offspring. If, on the other hand, a gold male is crossed with silver hens, a true sex-link is achieved in that male offspring will be silver and female offspring gold. The Light Sussex is genetically silver and the Red Sussex is genetically gold.

Since the White Sussex is perhaps the least common of all Sussex colours, and the least manipulated by the show scene or modern utility breeding, it is likely that it is the only colour that still represents the true type and utility qualities of the breed. For this reason it could certainly rate as a highly useful and productive variety for the poultry keeper, either in its standard form or when crossed. The plumage should be pure white throughout.

Brown and Coronation Sussex are the rarest Sussex colours as of 2011. Brown Sussex males are in short supply, and along with the hens, have some recent Red Dorking and Red Sussex ancestry, the main purpose of such crossing being to preserve and continue the existence of the variety. If Brown Sussex cockerels are not available some breeders utilize a Red Dorking or Red Sussex cockerel as an alternative. The Red Dorking is likely to be an original parent

Large White Sussex cock.

Large White Sussex pullet.

Large Red Sussex hen.

Large Red Sussex cock.

breed of the Sussex, and has the same distinctive black-red coloured males of the Brown Sussex. The most obvious difference between them is that the Red Dorking has five toes while the Sussex has only four.

The Brown Sussex has always had a close association with the Red variety, so crossing with the Red Sussex is one of the safest options to retain type and possibly improve utility merit. However, the crossing of these two colours will often result in brown males exhibiting red on the breast and underbody, where it should be pure black. Such Brown Sussex males can be used and will yield

Large Brown Sussex hen.

Bantam Coronation Sussex female.

Two bantam Coronation Sussex females.

Day-old large Brown Sussex.

brown females, but careful selection of future males is required in order to maintain the correct black-red colour of Brown Sussex males.

There is a great deal of variation in the colour of Red Sussex chicks from any source, and colours from dark buff through red to red/black are seen. This is likely to have been caused by the accidental or deliberate introduction of another breed while abroad in Europe, though fortunately the chicks always develop into Red Sussex. Mature female stock may become mottled or mealy in colour with age, possibly due to being kept outside in the sun. Strive for a consistent red without dark peppering in both sexes, and to prevent damage to plumage colour ensure that birds have plenty of shelter from the sun.

Red Sussex should have a slate under-colour, though often this is not the case and most will develop a reddish-buff under-colour and may only exhibit some slate colour on small areas of the body, which usually fails to persist the full length of the feather. When selecting youngstock for future breeding a consistent, rich, dark

Large Brown Sussex cockerel. Ideally the breast and thighs should be solid black. Due to past crossings with the Red Sussex to improve the number of Browns the breast and thigh colour may be patchy.

red top colour is important, but make the selection of birds with slate under-colour a priority. This is an important issue which needs to be rectified in the variety. Obviously if you wish to maintain a productive breed do not forget to select for utility merit, size and type at the same time.

The Coronation Sussex is an unstable colour form due to the blue gene. A mating of a Coronation Sussex male with Coronation females will result in 50 per cent Coronation offspring, 25 per cent White Sussex offspring and 25 per cent Light Sussex offspring.

Slate undercolour can be found in Brown and Red Sussex, and to some degree in the speckled variety.

Slate undercolour on a Brown Sussex; the slate colour should persist to the base of each feather.

Main Uses

- **As a dual-purpose breed:** The Sussex is one of the most productive all-round native breeds. It is more productive than any British light breed, although it also has a larger appetite. It is hardy, but not athletic or as determined a forager as many of the light breeds
- **As a sitter (broody):** It is a reliable sitter, and a careful and defensive mother
- **For exhibition:** It is a popular breed for exhibiting, although it was not specifically developed for this purpose
- **For free-range rearing:** The most productive use for the Sussex today is as a free-range domestic egg producer
- **For crossing:** Sex-linked crosses allow for the sex of day-old chicks to be determined by their down colour. Crossings could also be made with some lightweight breeds to create a hybrid of high productivity and vigour with a lower appetite

Head of a large Light Sussex hen showing the desirable and traditional small beard that occurs below the beak between the two wattles.

4 Domestic Fowl: Auto-Sexing Breeds

The first auto-sexing breeds were created by Professor R. C. Punnett and Mr Michael Peace of the Cambridge University Agricultural Research Department. They were developed at the request of poultry breeders for a single breed of domestic fowl that produced different coloured male and female chicks. Previous to the development of the auto-sexing breeds only certain barred breeds had any auto-sexing ability, and this was not necessarily very reliable. Alternatively breeders were using sex linkage to yield male and female chicks of different colours for easy identification as day olds. Unfortunately this meant maintaining two different breeds, for example a genetically 'gold' breed such as the Rhode Island Red, and a 'silver' breed such as the Light Sussex. It was necessary to cross a 'gold' cockerel to 'silver' hens to achieve the true sex linkage, whereby female offspring were coloured like the male, and male offspring were coloured like the female (whereas if a 'silver' cockerel were used with 'gold' hens, all the offspring would be coloured like the cockerel, as 'silver' is genetically dominant).

A unique and individual breed that bred true would do away with the need for more than one breed, and would ultimately reduce the number of breeding pens required. There was also the general belief that pure breeds were better than hybrids, particularly as the show scene required breeds to meet a standard. According to David Scrivener, the Cambridge team were investigating the matter from 1922, and discovered the auto-sexing principle by accident when experimentally crossing Barred Plymouth Rocks with Campines. This later became the Cambar breed, first shown at the World Poultry Congress in 1930, though the original breed is now thought to be extinct. The Cambar did not look much like the Campine, but all following auto-sexing breeds developed by the team were selected to maintain the characteristics of the other breed used.

To create an auto-sexing breed, a utility strain of Barred Plymouth Rock was used to establish the barring gene in a breed of commercial importance – for example the Leghorn, Sussex or Rhode Island Red. The resulting auto-sexing breed was therefore made to look like a different colour version of the Leghorn, Sussex or Rhode Island

Rhodebar chicks, male (left) and female (right). Autosexing ability is not 100 per cent accurate and is not always as distinctive as in this photo.

Red. The barring gene is sex linked and dominant. Males can carry two copies of the barring gene, whereas females can only carry one. Females are always one shade, generally with equal width white and black bars. Males that have only one copy of the barring gene will appear the same tone as females, and will often have the occasional solid-coloured feather. Males with two copies of the barring gene appear much paler, with wider white bands. These are genetically pure for the barring gene, and when mated to barred females will produce 100 per cent barred offspring. Males with one barring gene will produce as much as 25 per cent self-coloured offspring. In the case of the Rhodebar they would be Rhode Island Red.

The auto-sexing breeds had the potential to become highly useful commercial breeds, but unfortunately they were developed only shortly before the time of the specialized commercial hybrid after World War II. These grew more efficiently and were more productive. Today the auto-sexing breeds are largely forgotten by hobbyists, with the exception of the Cream Legbar, merely a selection of rare curiosities for just a few dedicated hobbyists. Perhaps if they were named after a local area or after Cambridge they may have had a greater following. Nevertheless, despite the lack of interest in the fancy for auto-sexing breeds, potentially they remain the most useful and adaptable of all native poultry breeds.

BRUSSBAR

Origin: Cambridgeshire, England
Class: Heavy, soft feather
Colour varieties: Gold, silver
Purpose: Meat, eggs
Eggs: Tinted
Weight: Large fowl cock, 4.1kg (9lb); cockerel, 3.6kg (8lb); large fowl hen, 3.2kg (7lb); pullet, 2.7kg (6lb)
Ancestry: Brown Sussex, Light Sussex, Barred Plymouth Rock
Sitter: Yes
Auto-sexing: Yes

History

Year created: Standardized by the Poultry Club of Great Britain in October 1952.

Originator: Professor R. C. Punnett and Michael Peace.

Breed development: The Brussbar was developed as an auto-sexing version of the utility Sussex breed. It was a logical step for the Cambridge research team, as the Sussex was the most popular commercial breed of Great Britain in the first half of the twentieth century, and if a breed with equal merit and close affinity with the Sussex could be created with auto-sexing ability it could prove to be an efficient commercial alternative. Punnett and Peace needed to use the Brown Sussex as they could not see how to produce the desired result from the Light Sussex. The Brown Sussex offers the initial sex link required. The Brown Sussex is purely an exhibition variety and has poor utility merit, so the commercially popular Light Sussex was later used to improve the utility merit of the Brussbar. (D. Scrivener, 2006.)

Only the gold variety currently exists, and is maintained by Mr Andrew Sheppy of

Congresbury, Bristol. He has been prudent enough to preserve the breed since the 1970s, at a time when many British breeds were fast diminishing. Unfortunately as of 2011, due to many fox attacks Andrew has only two young cockerels and two old hens remaining.

It is unlikely that anyone else will have the opportunity to breed and maintain the original Brussbar, but it could be recreated using Barred Plymouth Rocks, Brown Sussex and Light Sussex breeds.

Bantam: There has never been a bantam version.

Breeding and Management

The Brussbar was developed to be an auto-sexing version of the highly productive dual-purpose Sussex breed. They were never popular, commercially or domestically, despite the Brussbar having the potential to be one of the most useful native breeds. Unfortunately by the time the Brussbar was standardized, the broiler industry was beginning to expand and the pure breeds in general became largely redundant. White broilers that plucked out without any dark stubs were most desirable. The Brussbar, which is now the rarest native breed, did not have any chance of becoming useful in the poultry industry due to the specific development of table or laying hybrids. Today if they were in abundance they would be an excellent choice for the hobbyist requiring a dual-purpose breed.

Fortunately a new strain could be created because the original breeds used in the development of the Brussbar still exist. The trouble will be in finding reliable utility strains to make the process worthwhile.

Main Uses

- **For auto-sexing:** Allows cockerels to be identified and removed at day old
- **As a dual-purpose breed**
- **For free-range rearing**

Brussbar cock belonging to Andrew Sheppy of the National Poultry Collection. (Courtesy Andrew Sheppy, the Cobthorn Trust)

LEGBAR

Origin: Cambridge, England
Class: Light, soft feather
Colour varieties: Gold, silver, cream
Purpose: Eggs
Eggs: White or cream (Gold and Silver Legbars); blue, green or olive (Cream Legbar)
Weight: Large fowl cock, 2.7–3.4kg (6–7.5lb); large fowl hen, 2–2.7kg (4.5–6lb)
Ancestry: Brown Leghorn, Barred Plymouth Rock
Sitter: No
Auto-sexing: Yes

History

Year created: The Gold Legbar was standardized in January 1945, the Silver Legbar in December 1951 and the Cream Legbar in May 1958.

Large Gold Legbar pullet: this is the original variety of Legbar and is particularly rare.

Originator: Professor R. C. Punnett and Michael Peace.

Breed development: The Legbar is the second auto-sexing breed to be created by Professor R. C. Punnett and Michael Peace (the first being the Cambar, of which the original breed is believed to be extinct). It was created to make a commercially useful auto-sexing breed with the excellent laying abilities of the Leghorn.

The Gold Legbar was created by crossing a Brown Leghorn cock with Barred Plymouth Rock hens, followed by the selection of barred offspring and crossing these back to the Brown Leghorn. The brothers and sisters of the following generation were mated together, and only those male chicks that were genetically pure for the barring gene (the palest barred chicks) and barred females were kept and bred together to create the Gold Legbar breed. The Gold Legbar still exists, but the silver is likely to be extinct. The Silver Legbar was developed using the Gold Legbar and Silver Cambar.

The Cream Legbar was developed using the Araucana, a breed from South America that produces blue-coloured eggs. The crest was retained in the Cream Legbar to allow easy identification between it and the original Gold and Silver Legbars. (D. Scrivener, 2006.)

Bantam: Bantams do exist but are very rare.

Breeding and Management

The Legbar is a hardy, very energetic but not particularly flighty breed primarily used for domestic egg production. The Gold Legbar produces large white eggs, the Cream Legbar large eggs of various shades from pale green to blue, depending on the strain. The Gold Legbar is almost extinct and very few people appear to be showing an interest in this variety at present, despite

Head of a Gold Legbar pullet: the single comb can fall over to one side of the face in some hens, but it is objectionable if the eyesight is blocked.

it potentially being the most productive white egg layer that exists among native breeds.

On the other hand, the Cream Legbar – also known as the Crested Legbar – is very fashionable and numerous because it produces blue eggs. It is one of very few breeds that can do this, and owes this ability to the Araucana; indeed the Cream Legbar may well exceed the productivity and egg size of the Araucana, though it is essential to locate pure stock with a crest and correct body type. Plumage markings can vary depending on the strain and it is important to study the breed standard carefully to ensure that correctly coloured stock is sourced. Egg colour and depth of colour vary greatly, and even if a desirable deep blue egg layer is found, there is no guarantee that its egg-laying ability will also be high. For those keen to develop a commercially viable blue egg layer, the Cream Legbar would be the breed to start with.

The Legbar has a large single comb that can be vulnerable to frost damage during the winter. To avoid any such damage, ensure birds have continued access to a dry and insulated house, and if necessary use Vaseline on the combs to offer some protection.

The leg colour should be yellow, although birds with white legs are often seen: this may be an indication of past crossing.

To prevent continued inbreeding, further lines can be created by crossing a Gold Legbar male with Brown Leghorn females, then keeping the barred females and crossing them back with a genetically dominant barred male. Barred males are either a light or dark shade, and those that are palest have two barring genes and are the sort required for breeding. Try to source a utility strain of Brown Leghorn wherever possible to make the breeding effort worthwhile.

Main Uses

- **For auto-sexing:** Allows cockerels to be identified and removed at day old
- **As a productive egg layer:** The gold variety lays white eggs, the cream (crested) variety lays pale blue or green eggs

A breeding pen of Cream Legbar. (Wernlas collection)

RHODEBAR

Origin: Storrington, Sussex, England
Class: Heavy, soft feather
Colour varieties: Red-barred
Purpose: Eggs
Eggs: Brown
Weight: Large fowl cock, 3.85kg (8.5lb) min.; cockerel, 3.6kg (8lb); large fowl hen, 2.9kg (6.5lb) min.; pullet, 2.5kg (5.5lb)
Ancestry: Rhode Island Red, Barred Plymouth Rock
Sitter: Yes
Auto-sexing: Yes

History

Year created: The Rhodebar was standardized in October 1952.

Originator: There are different originators depending on the strain. Thus Mr B. de H. Pickard of Storrington, Sussex, developed a strain using the Barred Plymouth Rock and Rhode Island Red. Other breeders created Rhodebar strains using different methods. (D. Scrivener, 2006.)

A dark-barred bantam Rhodebar cock: it is genetically impure because it carries only one barring gene. This type should not be used for breeding, because such birds will throw a small percentage of pure Rhode Island Red offspring.

Breed development: The Rhodebar was developed as an auto-sexing breed by

Large Rhodebar hen. All Rhodebar hens are genetically impure with only one barring gene, meaning that they only come in one shade of red-barred.

Large Rhodebar cock: a genetically pure male with two barring genes, of the pale sort that is required for breeding pure Rhodebar.

Bantam Rhodebar hen.

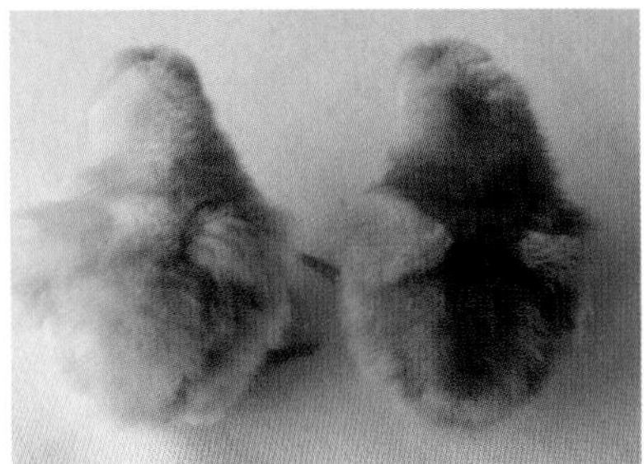

Day-old Rhodebars. The pale-coloured chick to the left is the cockerel, the dark buff chick to the right is the pullet.

outcrossing to the commercially popular American Rhode Island Red for the latter's egg-laying abilities. Until the 1960s the Rhode Island Red was one of the most popular commercial poultry breeds used in Great Britain, so clearly an auto-sexing version could have been highly lucrative.

Bantam: The bantam variety was created by Brian Sands of Friskney Eaudyke, Boston, Lincolnshire. It is rarer than the large fowl.

Breeding and Management

The auto-sexing version of the Rhodebar was developed to be an excellent egg-laying breed where cockerels could be easily identified and removed early on from the growing stock. The Rhodebar was based on the Rhode Island Red, and retains its type and utility merits. Rhodebars could be improved or reinvigorated by reintroducing utility strains of the Rhode Island Red.

The Rhodebar is neither flighty nor aggressive. It is as wary as the Sussex and not quite as easily handled as the Orpington, but will adapt as happily to free-range rearing as to being kept in an ark or aviary. Unfortunately it came into existence at the end of the pure breed era, at a time when specific hybrid stock was being developed for egg farms.

The Rhodebar could certainly prove to be an excellent choice for those wanting a small free-range egg-laying flock, but who don't have the space to grow on all birds in order to select and determine sex: the auto-sexing ability means that unwanted cockerels can be disposed of early on, thereby minimizing demands on space and feed requirements, and noise nuisance. The bantam version is also a productive layer of pale brown eggs, but may be better used for exhibiting or for those who wish to keep poultry but have very little room to do so.

To prevent extensive inbreeding, further lines can be created by crossing a Rhodebar male with Rhode Island Red females, then keeping the barred females to be crossed back with a genetically dominant barred male. Barred males are either a light or dark shade: the palest have two barring genes and are the sort required for breeding to prevent pure Rhode Island Reds coming out in the offspring.

Main Uses

- **For auto-sexing:** Cockerels can be identified and removed at day old
- **As a dual-purpose breed:** The Rhodebar is potentially one of our most useful and productive poultry breeds: it is primarily developed as an egg producer but also has table value
- **In high density residential areas:** The bantam version can be kept in small numbers in confined pens

WELBAR

Origin: Eastwrey, Lustleigh, Devon, England
Class: Light, soft feather
Colour varieties: Gold, silver
Purpose: Eggs
Eggs: Brown
Weight: Cock, 2.95–3.4kg (6.5–7.5lb); hen, 2.25–2.7kg (5–6lb)
Ancestry: Welsummer, Barred Plymouth Rock
Sitter: No
Auto-sexing: Yes

History

Year created: Developed in the early 1940s and standardized in October 1948.

Originator: Mr H. R. S. Humphreys.

Breed development: The Welbar was developed from the Dutch Welsummer breed, the dark brown eggs of the Welsummer being popular among hobbyists. The silver and gold varieties were developed simultaneously, but the gold is the only colour now commonly seen. (D. Scrivener, 2006.)

Large Welbar male.

Large Welbar hen.

Bantam: Created by John Buck of Bristol. (D. Scrivener, 2006.)

Breeding and Management

The Welbar could easily become one of the most popular of the auto-sexing breeds, largely due to the reasonable quantities of medium-sized, dark brown eggs that it produces: it lays the darkest (terracotta) eggs of any native breed.

Like the Welsummer, it is an active and hardy lightweight breed suited to free-range rearing. It is not particularly flighty. The bantam variety might be ideal for those who wish to have small to medium-sized dark brown eggs but who don't have enough room to keep large birds. Similarly breeders with only room for one or two small aviaries or arks would be well advised to use the bantam Welbar, as the auto-sexing ability is useful for identifying and weeding out the cockerels early on, before they grow and take up room or become noisy.

To prevent extensive inbreeding, further lines can be created by crossing a Welbar male with Welsummer females, then keeping the barred females and crossing them back with a genetically dominant barred male. Barred males are either a light or a dark shade, but the ones required for breeding are the palest because these have two dominant barring genes.

Main Uses

- **For auto-sexing:** Cockerels can be identified and removed at day old
- **As a productive egg layer:** It is one of few native breeds to lay brown eggs
- **For free-range rearing:** It is hardy and energetic
- **In high density residential areas:** The bantam version can adapt to being kept in small numbers in confined pens

WYBAR

Origin: Essex
Class: Heavy, soft feather
Colour varieties: Gold, silver
Purpose: Eggs
Eggs: Tinted
Weight: Cock, 3–4.1kg (6.5–9lb); hen, 2.5–3.2kg (5.5–7lb)
Ancestry: Gold and Silver-laced Wyandotte, Barred Plymouth Rock, Brussbar
Sitter: Yes
Auto-sexing: Yes

History

Year created: 1950–51.

Originator: Lt Commander A. O. Foden.

Breed development: The Wybar was developed in Hadleigh, Essex (D. Scrivener, 2006.) and is an enhancement of the practical qualities of the Wyandotte, allowing for sexing as day olds. The Wyandotte is an attractive American breed popular with exhibitors, and as a useful dual-purpose utility breed for domestic use, although it has never been as productive and commercially viable as the Sussex or Rhode Island Red.

Bantam: Bantam Wybar are more common than the large fowl, and currently exist in both silver and gold varieties.

Breeding and Management

The Wybar was developed as a triple-purpose breed, for eggs, meat and exhibition, with auto-sexing an added benefit. It is based on the American Wyandotte, but although productive, is not as useful as the auto-sexing breeds based on Sussex, Rhode Island Red or Leghorn. They are suitable for free-range rearing and are now predominantly used for exhibiting or as attractive pets.

The bantam can only truly be used as an exhibition bird or for those with limited space. Like the bantam Rhodebar and bantam Welbar, they can be maintained in small pens or aviaries. Their auto-sexing ability allows for the removal of cockerels early on before they grow and take up valuable space, and before they become a noise nuisance.

One benefit of using Wybars is that new bloodlines can be created easily, by crossing a Wybar male to Silver-laced or Gold-laced Wyandotte females, and then selecting the barred females and crossing back to a genetically pure barred male. The palest

LEFT: Bantam Silver Wybar male with correct square-fronted rosecomb, with fine workings and single leader following the curve of the neck.

RIGHT: Bantam Silver Wybar female.

LEFT: Bantam Silver Wybar female.

RIGHT: A bantam gold Wybar male.

males have two barring genes and are the ones that should be kept for breeding; all dark-barred males should be discarded. If males with only a single dose of the barring gene – that is, the darker ones – were kept for breeding, the original Silver-laced or Gold-laced Wyandotte will reappear in the offspring.

Wybars have a rosecomb that is evenly covered with 'rounded joints'. The comb should taper gradually to the leader, which should curve downwards, following the neck line. The comb is square-fronted and held closely over the head, but it must not obstruct the vision.

Wybars do not require any special housing requirements, although confinement within a small sheltered pen would ensure their plumage remains in excellent condition for showing. The silver-white ground colour of the silver variety may develop brassiness – a yellowish tinge to white plumage – if birds are left to roam outdoors in full sun. Brassiness is not a desirable feature in exhibition birds.

The large fowl is extremely rare. The gold variety may no longer exist, but could be recreated.

Main Uses

- **For auto-sexing:** Cockerels can be identified and removed as day olds
- **As a dual-purpose breed**
- **In high density residential areas:** The bantam version can adapt to being kept in small numbers in a confined pen
- **For exhibition:** Attractive laced and barred plumage

Bantam Silver Wybar female.

5 Waterfowl: Domestic Ducks and Geese

Domestic Ducks

ABACOT RANGER

Origin: Colchester, Essex
Class: Light
Colour varieties: Silver
Purpose: Eggs
Eggs: White
Weight: Drake, 2.3–2.5kg (5–5.5lb); duck, 2–2.3kg (4.5–5lb)
Ancestry: Campbell, White Indian Runner
Sitter: No
Auto-sexing: No

History

Year of creation: About 1917.

Originator: Mr Oscar Grey of Friday Wood, Colchester, Essex.

Breed development: Originally but briefly known as the Hooded Ranger Duck, the Abacot Ranger was developed from white 'sports' thrown from Khaki Campbells, possibly due to an impurity in the Campbell stock. These were mated together and the resulting progeny crossed with a White Indian Runner drake. However, by the middle of the twentieth century the breed had virtually disappeared in Great Britain.

Herr Leiker, living in Germany, had acquired some Abacots at an earlier date, and by 1928 had stabilized the most common colour pattern. The Abacot Ranger, named

Abacot Ranger duck: the crown of the head has dark graining.

Abacot Ranger drake with a black head, with a green sheen and a white ring encircling the neck.

Leikers Streifere (Leikers Ranger) by Herr Leiker, was standardized in Germany in 1934. Stock was later brought back to the UK, and a translation of the German standard was proposed by the British Waterfowl Association in 1987; this was later accepted by the Poultry Club of Great Britain. (Chris and Mike Ashton, 2008.)

Breeding and Management

The Abacot Ranger is a small, alert and energetic breed with great egg-laying potential, well suited to back garden or orchard rearing. In many ways it is similar to the Campbell, but it offers to the hobbyist an alternative colour to the range offered by the Campbell.

Realistically, the Abacot Ranger has a lot of competition in that it is up against several well known specifically egg-producing breeds such as the Campbell and Welsh Harlequin, as well as the Silver Bantam or Miniature Silver Appleyard, which might well be used as low appetite egg producers. So unfortunately the breed is rarely used, and soon birds may well only be kept for exhibition. Currently they have a hardy and active habit and low appetite, making them useful for free range rearing.

The Abacot Ranger is similar in many ways to the Welsh Harlequin, both being of similar size and weight, both having Campbell parentage, and both with the same practical attributes. However, they differ in colour, namely the colour of the head in males, and the colour of the secondary feathers that form the speculum in both sexes. Thus in Abacot males the head is black with a green lustre, and the speculum in both males and females is blue-green with a black, then white bar at the tip of each feather. Welsh Harlequin drakes, on the other hand, have a dark brown head with a bronze-green lustre, and a bronze speculum with a green lustre and a white border at the tip of each feather.

Female Abacots have a more distinct fawn head colour than Welsh Harlequin ducks. The head of females is fawn with dark graining to the crown, with no eye stripes on the face. The fawn is a deeper colour in young females, fading with age. When young there should be a sharp divide between the fawn of the head and the cream-white of the upper breast.

The neck ring on male Abacots must not be absent, nor broken by another colour, nor excessively wide.

Main Uses

- **As a productive egg layer**
- **For low maintenance free-range rearing**
- **For exhibition**
- **For grazing and agro-forestry:** For use in new or established orchards and vineyards

Wing of an Abacot Ranger drake showing the required blue-green speculum with a black, then white bar near the tip of each feather that forms the speculum.

AYLESBURY

Origin: Aylesbury, Buckinghamshire, England
Class: Heavy
Colour varieties: White
Purpose: Meat, exhibition
Eggs: White
Weight: Drake, 4.5–5.4kg (10–12lb); duck, 4.1–5kg (9–11lb)
Ancestry: Not known
Sitter: Not often
Auto-sexing: No

History

Year of creation: Not known; it is one of the oldest native duck breeds.

Originator: Not known.

Breed development: As early as 1750 the poor of Aylesbury supported themselves by breeding young ducks. Furthermore a demand for table ducks in London during the eighteenth and nineteenth centuries created a blooming trade in Aylesbury for its white ducks. There were many large farms in the Vale of Aylesbury, in villages such as Haddenham, but small duckers who were labourers by trade were also able to rear as many as 400 to 1,000 ducks per year in their own homes. (Chris and Mike Ashton, 2008.)

Aylesbury duck close to the original commercial type with no keel and a tight, rounded breast, making the bird suitable for table use.

An Aylesbury drake of the exhibition type. The pink bill and head can reach up to 20cm (8in) long, and lines up almost level with the top of the head. The keel is horizontal and most distinctive when standing at rest.

Breeding and Management

The Aylesbury was a world-renowned, white-skinned table duck, however the number of original utility-type Aylesburys has hugely diminished and it is now primarily used as an exhibition breed – however, there is much confusion among domestic poultry keepers regarding the true identity of the Aylesbury. Many continue to breed a varied mix of ducks due to the lack of information and publicity describing the breed.

To understand clearly what an Aylesbury duck is, it is necessary to group the various forms that currently exist into the utility, the exhibition and the hybrid Aylesbury types.

Utility Aylesbury: The utility is the original Aylesbury used commercially for the production of table ducks for the London market. They have a more rounded breast and less prominent keel than the exhibition type. There is believed to be only one breeder of the original type, although a small number may still exist in the wider duck population – but their degree of purity is not known.

Exhibition Aylesbury: The exhibition version was an exaggeration of the utility Aylesbury, developed soon after the first poultry shows of the mid-twentieth century to exaggerate the typified Aylesbury table duck. The work of those that exhibited the breed soon made them a unique type among duck breeds, creating an exaggerated, long straight keel with a flap of skin that has no practical use. The eggs are large to extra large, and pure white.

Hybrid Aylesbury: The hybrid type has been developed by crossing with Pekin ducks that have been imported from China since 1873 (Chris and Mike Ashton, 2008) or by the amalgamation of other white duck breeds. This has partly been for the improvement of table qualities and growth rate, but also due to unscrupulous breeding by hobbyists. These birds often have a more upright habit, but will vary greatly in size depending on their parentage, and will usually have a yellow or orange bill.

During the twentieth century the demand for Aylesbury ducks diminished, and the long-established flocks were sold up one by one. The few remaining true commercial Aylesburys now belong to Mr Richard Waller of Chesham, Buckinghamshire, whose family was prudent enough to continue to keep and preserve the Aylesbury at a time when no one else was truly interested. However, to assume that Richard is the only keeper of the true Aylesbury would be a bold statement, because occasionally utility Aylesburys can be found, although their level of purity is not known.

Richard Waller's flock is known to consist of pure authentic Aylesburys, and he is now probably the only known keeper and breeder of the utility type. However, because his stock is only sold as plucked table birds, the Aylesbury duck name is now used incorrectly to describe any large white utility duck, with countless poultry keepers using the term 'Aylesbury' indiscriminately, due to the lack of an easily available utility Aylesbury type. With the utility Aylesbury all but extinct, fanciers have turned to the exhibition Aylesbury for comfort: this type has been developed over a very long period and is still maintained by a few keen breeders.

It is very hard to source reliable stock, and even some exhibited strains have been influenced by past crossing. Any sign of yellow or orange in the beak is a sure sign of past cross-breeding. Hybrid Aylesburys are often highly productive and very hardy, and such white mongrel farmyard ducks can currently be purchased for next to nothing at auctions.

Aylesburys used for breeding should be kept at a ratio of one drake to three or five ducks. Select a light, energetic drake to improve fertility. Despite common belief, a pond is not essential for fertility, although it would help these heavy ducks in the mating process.

Exhibition Aylesburys should have clean, pure, satin white plumage, a long, horizontal and straight keel, and a long and broad body with a flat back. The bill should be flesh pink without blemishes, and should line up nearly flush with the top of the head. The head and bill of standard-bred exhibition Aylesburys measure from 15–20cm (6–8in) long when they reach maturity.

One of several variations of Aylesbury hybrids, the carriage of this drake is midway between a Pekin and an exhibition Aylesbury. The head is very large, like a Pekin, and although the bill and topline are similar to that of an Aylesbury, the colour is yellow.

To maintain exhibition Aylesburys in prime condition for showing, ensure they have a large grass run and a pond to keep fit and clean during the spring and summer months. In winter keep birds in sheds with straw bedding, and provide access to a secure outdoor area of hard standing for birds to maintain condition. Access to the elements and plenty of water will prevent problems such as wet feather, which often occurs in dusty environments. A veranda with a floor of 2 × 2in weldmesh raised above ground would also make clean and manageable outdoor access.

Aylesburys are not flighty and do not forage as eagerly as the many native lightweight breeds, but they could still turn a pasture into a mud bath in winter, given time.

Main Uses

- **For exhibition**
- **As a table bird**
- **For their down for duvets and pillows:** Aylesburys and Aylesbury-Pekin crosses have pure white plumage throughout

An Aylesbury hybrid duck with a short pink bill and a carriage nearer to that of the Pekin.

Head of a correct Aylesbury female. The top of the upper mandible lines up nearly flush with the top of the head. The bill is pinky white, and the eye is blue and held near to the top of the skull.

CAMPBELL

Origin: Uley, Gloucestershire, England
Class: Light
Colour varieties: Dark, khaki, white
Purpose: Eggs
Eggs: White; green in dark variety
Weight: Drake, 2.3–2.5kg (5–5.5lb); duck, 2–2.3kg (4.5–5lb)
Ancestry: Rouen, Fawn and White Indian Runner and Mallard
Sitter: No
Auto-sexing: No

History

Year of creation: Khaki 1901; White 1924; Dark first mentioned in 1943.

Originator: Mrs Adele Campbell of Uley in Gloucestershire originated the Khaki. The White Campbell was developed by Captain F. S. Pardoe. Mr H. R. S. Humphreys of Lustleigh, Devon, created the Dark Campbell.

Breed development: In the first instance Mrs Campbell crossed a Fawn and White Indian Runner Duck with a Rouen drake. At some point wild mallard was introduced into the breed. The progeny were later developed into the Khaki Campbell. The White Campbell was developed with similar egg-laying ability to the khaki and with the potential to be used in crosses to produce white table ducks for commercial use. (Chris and Mike Ashton, 2008.)

White feathering is preferred as no dark stubs are left on the carcass once plucked.

The dark variety was created to make sex-linkage in ducks possible.

Khaki Campbell duck.

Khaki Campbell drake; the carriage in Campbells should be held at about 35 degrees.

Breeding and Management

Campbells are a very active and energetic breed with a slightly nervous disposition. They are not flighty, but are quick on their feet and hardy. They are specifically an egg producer, being potentially the most productive layer of any duck breed. Eggs are medium sized and pearly white.

Continuing selection for high egg-laying ability is important in order to maintain and improve the main attribute of the breed. Campbells are energetic foragers so need to be permanently housed over winter if

Dark Campbell drake. The bill of Campbells joins smoothly with the top of the head.

the site they inhabit is particularly wet so as to prevent excessive wear and tear to the grazing area.

The khaki is the most common variety, the productivity of which depends greatly on the strain and source of the stock. Dark and white varieties still exist but are very rarely seen. The whites have undoubtedly been crossed in the past with breeds such as the Pekin to produce table birds. Some of these White Campbell crosses are now amalgamated with the general white mongrel duck of British farmyards, incorrectly but affectionately known as Aylesburys. If you wish to source White Campbells, study photos carefully and if necessary study the breed standard to help you select the correct stock rather than a mongrel. The White Campbell has the same carriage as the khaki; the differences are in colour: White Campbells have an orange bill and orange legs and feet, grey-blue eyes and pure white plumage throughout.

The Campbell would make an ideal breed for crossing with other small breeds to create a low maintenance, low appetite hybrid with high egg production.

In all colours there should not be a neck ring on drakes or ducks, and ducks should not have eye stripes.

Dark Campbell duck.

Main Uses

- **As a productive egg layer:** Traditionally the most productive egg layer of any of the native poultry breeds
- **For low maintenance free-range rearing**
- **For grazing and agro-forestry:** For use in new or established orchards and vineyards

White Campbell drake.

MAGPIE

Origin: Wales
Class: Light
Colour varieties: Black and white, blue and white, chocolate and white, dun and white
Purpose: Eggs, meat, exhibition
Eggs: Pale green
Weight: Drake, 2.5–3.2kg (5.5–7lb); duck, 2–2.7kg (4.5–6lb)
Ancestry: Not known
Sitter: Not often
Auto-sexing: No

History

Year of creation: Standardized in 1926.

Originator: Rev. M. C. Gower Williams and later Mr Oliver Drake.

Breed development: The Magpie was developed in Wales along scientific lines. The breed gets its name from its black and white pied colour markings similar to the wild magpie, and may have been created from the crossing of a black breed with a white one, with patience and careful selection. The actual parentage and method of development is not known. (Chris and Mike Ashton, 2008.)

Breeding and Management

Magpies are very hardy and keen foragers. They are not as energetic as Campbells or Shetlands, and are not remotely flighty. A triple-purpose breed, Magpies are suitable for free-range rearing to produce eggs and table birds, and with careful selection can be exhibited at shows.

The white-feathered breast and under-body ensure a clean appearance when plucked for the table. An attractive feature of the Magpie is the variation in markings that can occur in ducklings; this means that poorly marked birds if used for breeding can still yield correctly marked offspring. In the past, incorrectly marked birds have been used in the creation of the Ancona breed in the USA. The original colours of the Magpie were black and white, blue and white, and dun and white; however the black and white is the only colour commonly seen today, although blue and white and chocolate and white still exist today, but are rarely seen.

Black and White Magpie duck with preferred colour distribution.

An incorrectly marked Blue and White Magpie drake.

A correctly marked Chocolate and White Magpie drake.

Day-old ducklings are attractively marked and give a good indication of how they will be marked at maturity; this means that poorly marked birds can be disposed of early on. Adult Black and White Magpies should be predominantly white, with black on the back from the shoulders to the tail forming a heart shape when the wings are closed and viewed from above. There is a separate black marking on the crown of the head. The junction where the black plumage finishes and meets the white should be clean and sharply defined.

Birds to be exhibited must have correct, symmetrical markings. There should not be any other markings other than those described; there should not be any black on the face. Dark mottling on the legs is not desirable, but can sometimes be seen.

Magpies do not have any special housing requirements but would benefit from plenty of secure grazing and a simple but secure house. A pool for swimming and to maintain condition would be beneficial but is not essential. If birds are to be exhibited a pool would be very helpful to ensure they keep their plumage clean.

Main Uses

- **For exhibition**
- **As a dual-purpose breed:** For eggs and meat
- **For grazing and agro-forestry:** For use in new or established orchards and vineyards

One of the incorrectly marked birds that can occur, the black markings extending down the neck and on to the breast and thighs.

A Black and White Magpie duck. When the wings are closed the white of the primary wing feathers creates a black, heart-shaped marking on the back.

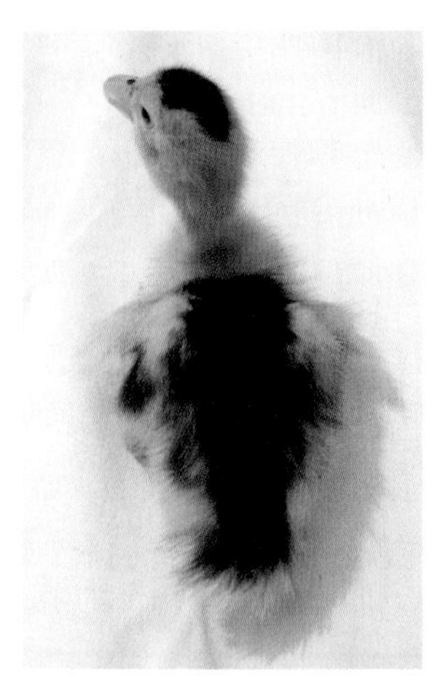

Black and White Magpie duckling.

The same duckling as in the previous photo, now nearly four weeks old with almost identical markings.

ORPINGTON

Origin: Orpington, Kent, England
Class: Light
Colour varieties: Buff, blue
Purpose: Eggs, meat
Eggs: White
Weight: Drake, 2.2–3.4kg (5–7.5lb); duck, 2.2–3.2kg (5–7lb)
Ancestry: Uncertain, possibly Aylesbury, Cayuga, Indian Runner, Rouen
Sitter: Not often
Auto-sexing: No

A Buff Orpington drake with a seal-brown head. The whole of the bird is the correct shade for a dark buff drake.

A Buff Orpington duck of the correct shade; small amounts of pencilling or a bluish rump may be evident in some dark buff specimens. This duck has undesirable pale eye stripes.

History

Year of creation: The Buff Orpington was standardized in 1910, and the blue in 1926. Both were advertised before 1900. A black variety was developed by 1913 and a chocolate variety by 1918, but neither has ever been standardized. (Chris and Mike Ashton, 2008.) As far as is known, the black and chocolate varieties no longer exist.

Originator: Mr William Cook of Orpington House, St Mary Cray, Orpington, Kent.

Breed development: The Orpington was developed as a dual-purpose duck breed, and to capitalize on the trend for buff-coloured plumage. William Cook was also responsible for the creation of the Orpington domestic fowl breed, including the buff variety.

Breeding and Management

The Orpington is an energetic breed suited to back garden and free-range rearing. They are hardy, keen foragers and are a productive dual-purpose breed. The eggs are medium to large in size, and pearly white. Buff and its variants is the most common variety; although Blue Orpingtons do still exist they are rarely seen. Buff plumage is easily damaged by the sun, so to keep plumage colour in prime condition birds would benefit from a sheltered pen – an orchard with shelter provided by trees would be ideal. An added benefit of grazing in an orchard is that the ground would remain drier in wet weather, preventing the ducks turning the grass and undergrowth to mud.

Over the winter months ducks are better off being kept indoors with access to an outdoor area of hard standing or a veranda with a weldmesh floor.

Buff Orpington ducks are impure for blue dilution, meaning that the desirable dark buff required of the breed standard is unstable. When standard-bred Buff Orpingtons are bred, three colour variants are produced. When a desirable dark buff drake is mated to a dark buff duck, both male and female offspring can be either pale buff, standard-bred dark buff, or khaki. To produce the desired dark buff required by the breed standard it is necessary to cross a pale buff with a khaki.

The blue dilution commonly results in pale buff birds having blue dilution to feathers in females, and a blue head in drakes. Blue dilution can sometimes be found in dark buff birds, especially on the rump of drakes and the wings of ducks, and is considered a fault. Eye stripes will occur in some female offspring: these are also undesirable and should be avoided in breeding stock.

Colour carries the greatest number of points at shows, and the dark buff variety is the type required by the breed standard. Pale buff and khaki offspring should not be shown.

Despite the potential confusion from having three colours from the offspring, the Buff Orpington remains the only native buff-coloured duck breed, and all the variants are equally useful from the utility egg-laying perspective. Also the buff plumage of the Orpington allows birds to be plucked clean without any undesirable dark-coloured stubs being left on the carcass.

Ideally the producer would keep a large flock of Buff Orpingtons with all three variants present: they would all be the same type and have the same egg-laying capacity, and it would be possible to select pale buff and khaki breeding stock to produce 100 per cent dark buff offspring either for sale, exhibiting or home table use.

A Blonde Orpington drake with a blue-grey head caused by the blue dilution which affects the breed.

A Blonde Orpington duck, with a pale buff shade throughout. Undesirable bluish feathers are most commonly seen in blonde specimens.

Main Uses

- **As a dual-purpose breed:** The Orpington has the potential to be the most productive and attractive native dual-purpose duck breed. Magpies and Stanbridge Whites are other alternatives
- **For grazing and agro-forestry:** For use in new or established orchards and vineyards

A brown duck with khaki-type pencilled markings.

SHETLAND

Origin: Shetland Isles, Scotland
Class: Light
Colour varieties: Black
Purpose: Eggs
Eggs: White
Weight: Drake, approx. 1.6kg (3.5lb); duck, approx. 1.4kg (3lb)
Ancestry: Uncertain
Sitter: No
Auto-sexing: No

History

Year of creation: Not known.

Originator: Not known.

Breed development: The history and development of the Shetland duck is largely unknown. It is possible that the ancestors of the breed were brought to the islands by Viking settlers, and similarities with breeds such as the Blue Swedish may support this claim. The existing Shetland was recovered by Mary and Thomas Isbister of Trondra, in the Shetland Isles. The first ducks they purchased had at some point been crossed with Khaki Campbells, which may explain the current tendency for some Shetlands to have brown undercolour and pencilling under the wing. Mary and Thomas later came across a pure female on the island of Foula, and a few others from the east side of Shetland. From these birds the Shetland breed was re-established. An exchange of drakes at a later date with a flock discovered on the north mainland of Scotland has helped prevent inbreeding.

White-bibbed Black Shetland. The ideal is black throughout with a green sheen except for a white bib. Most, however, have some brown undercolour and pencilling under the wing.

Breeding and Management

The Shetland is a hardy and durable breed with a voracious foraging habit, often resulting in a mud-covered head and a deterioration

Shetland duck with typical irregular breast bib markings.

in bill colour; this breed therefore benefits from roaming at free range on plenty of open grassland. They are one of the smallest and most energetic of the native breeds, and are usefully kept in a garden environment amongst shrubs and climbers, or in vineyards or orchards, as a means of eradicating snails and slugs and for weed control.

The Shetland duck would have supplemented its diet in the past by foraging on the poor marshy grazing of the Shetland Islands, and would have been helpful in consuming snails that play host to liver fluke, a parasite of grazing livestock such as sheep.

The Shetland stands semi-erect and has a lively carriage with a slightly nervous habit.

Much careful selection is still required in the Shetland breed to ensure consistent plumage colour. Both drakes and ducks are glossy black with a white bib. They do not usually have a green lustre all over, but it would be desirable if the lustre were consistent.

The ancestry of the Shetland is not known. The Black East Indian duck has the same

A brown Shetland derived from black, white-bibbed stock, a sign of impurity probably caused by the known past crossing with the Khaki Campbell.

A twenty-five-day old Shetland drake.

black plumage with a green lustre, black legs and bill, and is very similar in size and weight, so may be one of the parent breeds. The Black Swedish, which is derived from the breeding of the Blue Swedish, may also be involved. The Black and Splash Swedish are byproducts of breeding the unstable blue colour, but black and splash are not popular and are often disposed off or used in the creation of blues. However, it is quite possible that the Black Swedish, which has a white bib, was used in the creation of the Shetland. The Shetland was quite possibly created by selection alone of the Black Swedish, or by

Day-old Shetland duckling.

crossing with the smaller Black East Indian breed; nevertheless the habit and size of the Shetland is far from that of today's Swedish, suggesting that some substantial changes have occurred.

ABOVE: Black Swedish duck, front view. Swedish ducks usually require a very neat bib. This bird shows that irregularly marked bibs do occur on Swedish stock.

Main Uses

- **As a productive egg layer**
- **For low maintenance, free-range rearing**
- **For grazing and agro-forestry:** For use in new or established orchards and vineyards

LEFT: A Black Swedish duck, one of the resulting colours from breeding the unstable Blue Swedish. It is possible the Shetland was derived from the Black Swedish, but Swedish are considerably larger with a low carriage set at adout 25 degrees and a more refined bib. Leg and bill colours are very similar.

LEFT: Adult Shetland duck: its energetic foraging habit involving dibbing its beak in wet soil has worn away the black colour of the bill to reveal a grey-blue undercolour.

RIGHT: Young Shetland duck with predominantly black bill.

SILVER APPLEYARD

There are both large and miniature versions of this breed.

Origin: *Large:* Village of Ixworth, Suffolk, England
Miniature: Folly Farm, Bourton-on-the-Water, Gloucestershire
Class: Heavy
Colour varieties: Silver
Purpose: Eggs, meat
Eggs: White
Weight: *Large:* Drake, 3.6–4.1kg (8–9lb); duck, 3.2–3.6kg (7–8lb)
Miniature: Drake, 1.36kg (3lb); duck, 1.19kg (2.5lb)
Ancestry: Not known
Sitter: Yes (miniature)
Auto-sexing: No

Miniature Silver Appleyard drake. The black head has a green lustre, and silver-white markings can be seen on the cheek and on the brow of the eye.

History

Year of creation: Uncertain, perhaps some time shortly before 1947.

Large Silver Appleyard drake. A white ring 5–9mm (0.25–0.37in) wide should be seen completely encircling the neck on drakes.

Originator: Reginald Appleyard of Priory Waterfowl Farm, Ixworth, Suffolk.

Breed development: The Silver Appleyard was created some time in the mid-twentieth century; how and when is not certain, and Reginald Appleyard does not mention the creation of any of his own breeds in his own books. His guide *Ducks – Breeding, Rearing, Management*, published in 1949, makes no mention of the Silver Appleyard, although it is likely the breed existed prior to the publication. A painting of Silver Appleyards by Wippell was commissioned in 1947.

Shortly after their creation the Appleyard had nearly disappeared, perhaps due to a lack of publicity until the 1970s, when Tom Bartlett of Folly Farm, Bourton-on-the-Water, Gloucestershire, purchased some birds at Chipping Norton market that he recognized as Appleyards. Tom borrowed the Wippell painting from Reginald Appleyard's daughter, Mrs Noreen Godwin, and went about breeding and selecting his Appleyards to resemble the type and colour of those depicted in the painting. (Chris and Mike Ashton, 2008)

In Reginald's words the Silver Appleyard was: 'An effort to breed and make a beautiful breed of duck, combination of beauty, size and lots of big white eggs, white skin, deep long wide breasts.' (Chris and Mike Ashton, 2008.)

Bantam: Tom Bartlett later went on to create the miniature Silver Appleyard early in the 1980s by breeding down from the large Appleyards. (Chris and Mike Ashton, 2008.) The full history is not certain, and whether Call Ducks or Silver Bantams were used is not known.

Miniature Silver Appleyard duck.

Breeding and Management

The large Silver Appleyard is an active breed despite its size, but it is slower and more easily handled than most breeds. The Appleyard was developed as a dual-purpose breed, producing a carcass of meat larger than most breeds with a regular supply of large white eggs. Silver Appleyards are suitable for free-range rearing, but must have secure fencing and housing. Miniatures are sometimes flighty and more energetic than the large version. To prevent

Large Silver Appleyard duck with desirable restricted plumage colour.

Miniature Silver Appleyard duck of the slightly darker type than is required by the breed standard, with less restriction of colour on the face and body than necessary. Ducks should have fawn from the top of the head and along the back of the neck to the shoulder without any breaks in colour. Ideally this bird should have more creamy-white colour on the breast.

them flying over a neighbour's fence, their wings should be clipped or pinioned.

The miniatures are ideally suited for people with limited space, and are primarily a variety for exhibiting or for use as low appetite, low maintenance egg producers. Their eggs are medium sized and white. Miniatures do not need a pond but they can swim individually in a shallow bucket if space is tight. A small ark or aviary with a surface area amounting to 2–3sq m would be adequate for a pair or trio. Cats are not a threat to adult miniature Appleyards, but foxes are. They are flighty and quick enough to evade a fox, but a secure pen and house would nevertheless be beneficial.

The colour of the Silver Appleyard is due to the restricted mallard gene, one effect of which limits the amount of pigment on the face and body – though its effects do seem to vary. On males, an unbroken white ring 5–9mm (0.2–0.4in) wide completely encircles the neck, while females have a fawn line from the crown of the head, down the back of the neck to the shoulder, which must not be broken by the silver-white of the neck. Females also have a fawn eye-stripe. The breast and under-body are creamy white. Currently some female stock is darker in colour than is required by the breed standard; often the breast and under-body lack creamy-white plumage, resulting in ducks that are closer to the colour of the Rouen Clair, a French breed that hobbyists sometimes confuse with Appleyards.

A clear sign of crossing can be seen in the colour of ducklings. Silver Appleyard ducklings must have predominantly yellow down with a dark Mohican stripe on the top of the head, and a dark tail. Any ducklings exhibiting a dark eye-stripe or predominantly black markings on the back, like wild Mallard ducklings, are not pure Silver Appleyards.

The size of large Appleyards is important: some stock is undersized, so careful selection and sourcing of large, well muscled birds is important in order to retain the traditional table qualities of the breed. To encourage high fertility it is advisable to keep the breeding ratios small, with perhaps one drake to three ducks, and wherever possible provide swimming water to make mating easier for this large breed.

Main Uses

- **As a dual-purpose breed:** The large Silver Appleyard lays large white eggs and is one of the most productive traditional table breeds
- **For exhibition:** The main purpose of miniatures is to exhibit them
- **For low maintenance, free-range rearing:** Suitable for miniatures only
- **For small gardens and high density residential areas:** Provided they are kept in arks or netted pens, preferably with clipped or pinioned wings

SILVER BANTAM

Origin: Village of Ixworth, Suffolk, England
Class: Light
Colour varieties: Silver
Purpose: Eggs, exhibition
Eggs: White
Weight: Drake, 1.36kg (3lb); duck, 1.19kg (2.5lb)
Ancestry: Khaki Campbell, White Call
Sitter: Yes
Auto-sexing: No

History

Year of creation: 1940s.

Originator: Reginald Appleyard.

Breed development: The Silver Bantam was developed from the crossing of a small Khaki Campbell and a White Call. (Chris and Mike Ashton, 2008.) As far as is known, the breed was developed to be a bantam version of the large Silver Appleyard. It is not certain why Reginald Appleyard felt the need to create the

Mature Silver Bantam duck.

A young Silver Bantam duck; it lacks the well defined fawn head that is typical of mature stock.

Silver Bantam duck: its breast is correctly coloured with a creamy-white ground streaked with light brown. The brow and crown of the head are correctly marked with dark graining, although the required fawn colour of the head of this specimen is barely visible.

bantam version, or why he did not use the large fowl in the creation of the bantam. Perhaps he was intending to create a small utility duck that could produce a lot of eggs like the Campbell, but by being smaller have a smaller appetite. Similarities with the Silver Appleyard may have been a secondary consideration that could not be realized.

Breeding and Management

The Silver Bantam is very rare, being maintained by very few breeders; it is perhaps the second rarest British waterfowl breed, and has largely been forgotten. It is sometimes flighty, and is a low maintenance breed capable of finding much of its own food if given a large enough grassy pen, or secure orchard or suchlike environment.

Due to its small size it is less messy and destructive than most other native breeds. A pair or trio could be kept in a small ark or shed, with access to a shallow bucket of water to drink from and swim in. It produces medium-sized white eggs.

The Silver Bantam has never been popular, but historically is a breed worth preserving, as the original bantam version of the Silver Appleyard as developed by Reginald Appleyard – even though it is not in fact related to the Silver Appleyard. Instead it looks similar in colour to the Abacot Ranger. The males are similarly coloured to Silver Appleyard drakes, but have a solid black head with a green lustre. Female Silver Bantams are like the Abacot Ranger duck, having a fawn-coloured head with darker graining on the crown. To prevent birds escaping, their wings should be clipped or pinioned.

Main Uses

- **For exhibition**
- **For low maintenance, free-range rearing**
- **For small gardens and high density residential areas:** Provided they are confined to a netted pen or have had their wings clipped or pinioned

STANBRIDGE WHITE

Origin: Stanbridge Earls, Romsey, Hampshire
Class: Light
Colour varieties: White
Purpose: Eggs, meat
Eggs: Pale green
Weight: Drake, 2.5–3.2kg (5.5–7lb); duck, 2–2.7kg (4.5–6lb)
Ancestry: Magpie
Sitter: No
Auto-sexing: No

History

Year of creation: Standardized in about 1930.

Originator: Captain C. K. Greenway of Stanbridge Earls, Romsey, Hampshire.

Breed development: The breed was most likely developed from white specimens of the Magpie breed, of which Captain Greenway was a well known breeder. The Stanbridge White is a very similar type to the Magpie, and both breeds lay a pale green egg.

The Stanbridge White has never been a widely known breed among people in the fancy, and was thought to be extinct until 2007, when the president of the Rare Poultry Society identified the breed in a flock kept by a Mr William Osborne of Lydney, Gloucestershire – although it is likely that other Stanbridge Whites do exist in other flocks, mixed in with other white breeds such as the White Campbell and hybrid Aylesburys, and probably crossed with these. Nevertheless William's small flock was the first to be identified, and is probably the purest stock that exists.

William Osborne originally purchased his ducks from an elderly man in the village of Aylburton in Gloucestershire, the village where

A Stanbridge White duck with a similar carriage to the Magpie and Campbell breeds. The Stanbridge was originally derived from the Magpie. The Magpie and Campbell hold themselves at about 35 degrees, which is somewhere in between the upright stance of the Pekin and the horizontal carriage of the Aylesbury.

Head of a Stanbridge White. The bill is correctly coloured pink with some natural discoloration due to injury and wear and tear. The best specimens will exhibit a slightly dished bill.

a known poultry photographer, Mr Arthur Rice, is said to have lived, and who also kept the Stanbridge White. At the time the elderly man did not know what they were, and regarded them simply as 'farmyard ducks'.

Breeding and Management

The Stanbridge White does not stand out as being particularly unique among duck breeds, but it has the potential to be a highly useful dual-purpose breed. Although it can be used as a table bird, it is primarily kept for its novelty value as a prolific layer of large green eggs.

Stanbridge Whites are easily handled and relatively calm when compared with other native breeds, and are hardy. They are not flighty, but they do exhibit similar habits to the versatile Magpie breed. The best specimens have a slightly dish-shaped bill; the bill should be pink and of medium length. Stanbridge Whites stand more upright than the Aylesbury, but not so much as the Pekin.

Unfortunately due to their rarity, which is largely brought about by their common white colour and similar habit to other established breeds, they have at some point in the past been crossed with White Campbells. This is evidenced when some ostensibly Stanbridge-bred stock produce white eggs and/or have a yellow beak. In the coming years it is to be hoped that the breed will become purified, but it is hard to see it surviving when so many other white ducks exist, particularly when many people seem content to breed and maintain the typical mongrel white farmyard duck, including hybrids of Aylesbury, Pekin, Campbell and White Indian Runner.

Yet the Stanbridge White is one of few native breeds to lay pale green eggs in abundance, making it a distinct alternative to Aylesburys or White Campbells. Crossing back to Black and White Magpies or white offspring from Magpies could reinvigorate the breed at a later date if required.

If birds are to be exhibited, ensure that the plumage is clean and pure white. The bill should be free from dark blemishes. To maintain them in prime condition provide access to plenty of sheltered grazing with a pool for swimming.

The Stanbridge White is probably the rarest of our native duck breeds. Currently it is kept by only a very few, it is in need of careful selection, and is difficult to source.

Main Uses

- **As a dual-purpose breed**
- **For its down for duvets and pillows:** The Stanbridge White has pure white feathers and down
- **For grazing and agro-forestry:** For use in new or established orchards and vineyards

WELSH HARLEQUIN

Origin: Hertfordshire, England and Criccieth, Gwynedd, North Wales
Class: Light
Colour varieties: Harlequin
Purpose: Eggs
Eggs: White
Weight: Drake, 2.3–2.5kg (5–5.5lb); duck, 2–2.3kg (4.5–5lb)
Ancestry: Khaki Campbell
Sitter: No
Auto-sexing: No

History

Year of creation: 1949–50.

Originator: Group Captain Leslie Bonnet.

Breed development: In 1949 Group Captain Leslie Bonnet hatched two honey-coloured ducklings, a duck and a drake, from his flock of Khaki Campbells at his home in Flaunden, Hertfordshire. From these, a small flock of pale-coloured ducks was created: they were considered to be as prolific as Khaki Campbells, and were named Honey Campbells. The Bonnet family later moved to Criccieth in North Wales, and these Honey Campbells were further developed. In about 1950 Captain Bonnet took the decision to change the name to Welsh Harlequins. All existing Welsh Harlequins originate from a flock owned by Mr Eddie Grayson of Rufford, Ormskirk, Lancashire, who purchased Welsh Harlequins from Captain Bonnet in February 1963. The flock owned by Captain Bonnet was largely lost to a fox attack in 1968 and his strain could not be recovered. He crossed the remaining Welsh Harlequins with his Whaylesbury hybrids to create what he called the New Welsh Harlequin; however, this strain probably no longer exists. (Chris and Mike Ashton, 2008.)

Welsh Harlequin drake; the head in the drake is dark brown with a bronze-green lustre, and terminates just above the shoulder where a white ring completely encircles the neck. The bill is olive green with a black bean. (Courtesy Martin Baldwin)

Welsh Harlequin duck; the bill of the duck is slate, tinged with green. The wing speculum on both male and female is bronze. (Courtesy Martin Baldwin)

Breeding and Management

The Welsh Harlequin is a lively breed with a semi-erect carriage at near 35 degrees to the horizontal. In size and composition it is similar to the Campbell. The Welsh Harlequin is notable for its attractive colour and egg-laying capability; the eggs are medium-sized and white.

Distinctively the head of drakes is brown with a bronze-green lustre. The secondary wing feathers that form the speculum are bronze with a green lustre and a white edge at the tip. This differs from the Abacot Ranger, which has a black head with a green lustre and a blue-green speculum tipped with black, then white bars. The presence of a blue speculum in Welsh Harlequins is a fault, and would suggest past crossing with another breed but not necessarily the Abacot Ranger. The neck ring on drakes should completely encircle the neck, and must not be broken by another colour.

The Welsh Harlequin is one of several breeds, including the Abacot Ranger and Campbell, with very similar utility merits and of similar size and habit. A hobbyist's selection could be made in favour of a colour or place of origin.

Main Uses

- **As a productive egg layer**
- **For exhibition**
- **For low maintenance, free-range rearing**
- **For grazing and agro-forestry:** For use in new or established orchards and vineyards

Welsh Harlequin pair. (Courtesy Martin Baldwin)

Domestic Geese

BRECON BUFF

Origin: Brecon Beacons, Powys, Wales
Class: Medium
Colour varieties: Buff
Purpose: Meat
Eggs: White
Weight: Gander, 7.3–9.1kg (16–20lb); goose, 6.3–8.2kg (14–18lb)
Ancestry: Unknown, possibly West of England
Sitter: Yes
Auto-sexing: No

History

Year of creation: Between 1929 and 1934.

Originator: The Brecon Buff was created by Rhys Llewellyn of Swansea.

Breed development: In 1929 Rhys Llewellyn discovered a buff-coloured goose in a flock of grey-and-white geese while driving across the Brecon Beacons. Interested in creating a buff-coloured breed, Rhys went in search of a buff gander, but with no luck. He eventually decided to use a medium-sized white Embden-type gander instead. All the progeny from this crossing (the buff goose and the white gander) turned out grey, and from these he kept back a gander. Later he acquired two more buff females from other hill farms, and bred these with this young gander: this cross produced several buff goslings. A buff gander was selected and bred with his female buff geese, and by 1934, 100 per cent buff progeny were produced. (Chris Ashton, 2010.)

Breeding and Management

Regarded as a dual-purpose breed, the Brecon Buff is hardy and has good mothering instincts. It is the largest of the native goose breeds, and can be used as a table bird, even though it is a medium-sized goose and smaller than the continental Embden and Toulouse geese.

The Brecon Buff can survive outdoors in all weathers, but the buff colour is vulnerable to damage from direct sun. A sheltered pen with plenty of vegetation and trees would limit sun damage. A grass pen with running water such as a natural stream, or a large pond at least a metre deep and more than 5m in diameter, would be ideal for two or three geese, although a small pool cleaned and topped up regularly would suffice in smaller quarters.

Brecon Buffs would tolerate an exposed site as long as they were provided with a basic or improvised shelter such as straw bales or a brash house or a small log cabin. They are prey to foxes so reliable fencing or electric netting is required.

When sourcing Brecon Buffs be sure to study the parent stock. True Brecons are lighter in build than the American Buff, and their beak, legs and feet are pink rather than orange: any sign of yellow is a sign of past crossing with the American Buff. The Brecon Buff has a dual-lobed

Brecon Buff gander and geese, showing their dual-lobed paunch between their legs.

paunch between the legs. Ideally the lobes should be equally proportioned.

Main Uses

- **As a table bird**
- **As a sitter (broody)**
- **For low maintenance free-range rearing**
- **For grazing and agro-forestry:** They are suitable for use in established orchards.
- **For exhibition**

Pair of Brecon Buffs, gander in the foreground, goose in the background. Note the correctly coloured bright pink legs, webs and beak. These birds are owned by Mike and Anne Griffiths of Kent.

SHETLAND

Origin: Shetland Isles, Scotland
Class: Light
Colour varieties: Grey and white
Purpose: Meat
Eggs: White
Weight: Gander, 5.5–6.4kg (12–14lb); goose, 4.5–5.5kg (10–12lb)
Ancestry: Not known
Sitter: Yes
Auto-sexing: Yes

History

Year of creation: The history of the Shetland goose is shrouded in mystery: they may have existed on the islands for many centuries.

Originator: Not known, but traditionally bred by the crofter of the Shetland Islands.

Breed development: The history of the Shetland is largely unknown, but it is possible they may share a common ancestry with the West of England breed. The Shetland may have been brought to the Shetland Isles from the mainland, and has since evolved and been selected for increased hardiness.

Breeding and Management

The Shetland is a hardy breed that has evolved and been selected to survive in a harsh, windswept environment with minimal maintenance. Having evolved on the Shetland Isles they are likely to have built up hardiness to exposed sites and wet ground. Shetlands can be grazed outdoors all year round, but they will soon muddy a grass pen in winter, so winter accommodation in a shed or barn with some secure outdoor hard standing would be best.

They can be used for grazing conservation areas, or in vineyards or plantations where they are required to find most of their own food and if necessary consume unwanted slugs and snails. Shetlands were traditionally used to graze pastures to consume the snails

Shetland geese. (Courtesy Mary Isbister)

Shetland gander on the left, goose on the right, and goslings. (Courtesy Mary Isbister)

that were the host to liver fluke, a parasite of sheep.

Shetlands lay between fifteen and twenty extra large white eggs each year before going broody and sitting. They have reliable brooding instincts, which, before the advent of incubators, would have been a necessity if the breed were to reproduce itself successfully. A pond or stream is not essential, although they would benefit from some swimming water to help maintain condition. The main consideration should be that they have plenty of secure grazing.

The Shetland and West of England differ in size, the Shetland being smaller. The smaller size may well be a result of the breed evolving through natural selection to suit the windswept and harsh winter climate of the Shetland Isles. Shetlands have pink beak, legs and webs; West of Englands have an orange beak and orange or pink legs and webs. They are both auto-sexing and the plumage colour and distribution of colour is very similar in both breeds.

The Shetland has a single-lobed paunch between its legs; all other native breeds have a dual-lobed paunch.

Main Uses

- **As a table bird**
- **As a sitter (broody)**
- **For low maintenance, free-range rearing**
- **For grazing and agro-forestry:** The Shetland can be used in established orchards

Shetland geese. (Courtesy Mary Isbister)

WEST OF ENGLAND

Origin: Great Britain
Class: Medium
Colour varieties: Grey and white
Purpose: Meat
Eggs: White
Weight: Gander, 7.3–9.1kg (16–20lb); goose, 6.3–8.2kg (14–18lb)
Ancestry: Unknown
Sitter: Yes
Auto-sexing: Yes

History

Year of creation: Not known.

Originator: Not known.

Breed development: The West of England is likely to be an ancient breed of farmyard geese developed over many centuries to be auto-sexing. It is certain that the auto-sexing trait pre-dates the mid-nineteenth-century Victorian poultry shows. (Chris Ashton, 2010.)

Breeding and Management

Like most geese, the West of England is hardy, but they are also energetic and active. They have a light build with tight plumage capable of withstanding a harsh, wet and waterlogged environment. West of England geese also have good brooding instincts.

The eggs are large, pure white, and produced in larger numbers than Shetland Goose eggs. The breed is suited to low maintenance, free-range rearing, and as long as there is plenty of secure grazing, they could find most of their own food. Housing can be very simple, but does need to be secure as they are still prey to foxes and dogs. West of Englands are easy to handle, and they are not flighty.

An interesting example of the hardiness and adaptability of the West of England has been seen on the Cambridgeshire Fens. A trio of West of England geese once lived on a smallholding which was demolished due to poor management; the geese were abandoned. The smallholding site was adjacent to a dyke, and these geese have survived for over fourteen years in a wild state. They appear to keep close to water at all times and have therefore successfully evaded predators, although they have not been known to rear any young in the wild. They are said to be more than twenty years old.

These semi-feral geese rely on water for their safety, but in a domestic situation it is not essential for birds to have a pond or stream. A large container of water to allow them to clean their bill will suffice, although they would relish the opportunity to take a swim.

The current struggle with the West of England is in finding pure stock. The gander is pure white, sometimes with a little grey on the back or head, but unfortunately many other breeds also have white ganders, so it is easy to appreciate that crossing is likely to have

A pair of West of Englands that have been roaming wild near Ramsey in Cambridgeshire for over fourteen years.

LEFT: Pure white one-year-old West of England gander. CENTRE: A West of England gander showing a small amount of grey markings on the back and head. This is common and acceptable among ganders. RIGHT: West of England goose. White markings seen at the front of the head are acceptable and may increase with age. The breast and lower neck should be pure white. The grey plumage on the breast of this goose may be a sign of past crossing with Pilgrim geese.

occurred, perhaps with the Embden, Roman and Pilgrim breeds.

The West of England is an auto-sexing breed, allowing for the sex of goslings to be determined at day old by the colour of the down. Male goslings have yellow down, and females have yellow down with grey markings on the back and head. If the breed has lost any auto-sexing ability to the extent that some females turn out mainly white or with unusual markings for the breed, then past crossing is likely to have occurred.

The West of England and Brecon Buff both have a dual-lobed paunch between the legs; the two lobes should be even in size. The paunch develops with age and is not evident until birds are around one year old.

Main Uses

- **As a table bird**
- **As a sitter (broody)**
- **For low maintenance, free-range rearing**
- **For grazing and agro-forestry:** For use in established orchards

Day-old female West of England gosling.

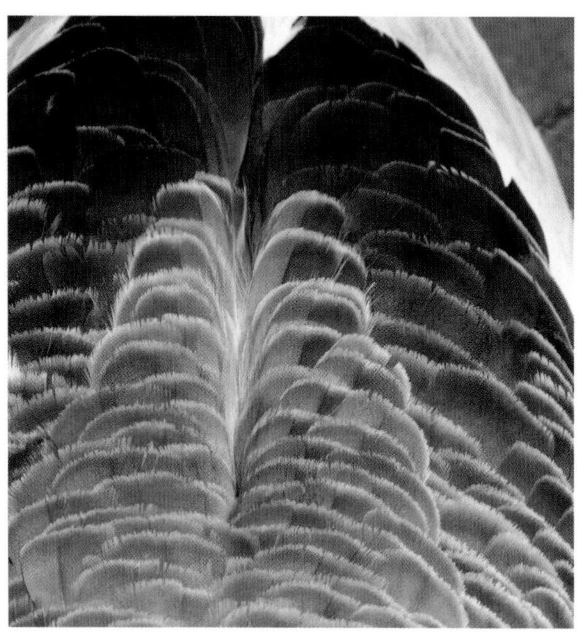

TOP: Blue eyes and orange beak are typical of the breed. The legs should be orange or pink.

RIGHT: The grey feathers are laced with a paler shade of grey.

BELOW: West of England goose showing the dual-lobed paunch between the legs which is characteristic of the breed.

Bibliography

Ashton, Chris, *Domestic Geese* (The Crowood Press, 2010).

Ashton, Chris and Mike, *The Domestic Duck* (The Crowood Press, 2008).

Atkinson, Herbert, *The Old English Game Fowl: Its history, description, management, breeding and feeding* (1891).

Roberts, Michael, D. L, and Brereton, Grant, *21st Century Poultry Breeding* (Gold Cockerel Books, 2008).

Roberts, Victoria, *British Poultry Standards, Sixth Edition* (Wiley-Blackwell, 2008).

Scrivener, David, *Popular Poultry Breeds* (The Crowood Press, 2009).

Scrivener, David, *Rare Poultry Breeds* (The Crowood Press, 2006).

Stevens, Lewis, *Genetics and Evolution of the Domestic Fowl* (Cambridge University Press, 1991).

Index